Coherence and Ultrashort Pulse Laser Emission

Coherence and Ultrashort Pulse Laser Emission

Editor

Gustaf Kron

scitus
academics

Coherence and Ultrashort Pulse Laser Emission

Edited by **Gustaf Kron**

ISBN: 978-1-68117-284-2
Library of Congress Control Number: 2016937455

SCITUS Academics LLC,
www.scitusacademics.com
Box No. 4766, 616 Corporate Way,
Suite 2, Valley Cottage,
NY 10989

Notice

Printed in United States of America on Acid Free Paper

Preface

More generally, coherence describes all the properties of the correlation between the physical quantities of one wave or several wave packets. The property of coherence is the basis for commercial applications such as holography, Sagnac gyroscope, radio antenna arrays, and opticalcoherence tomography and telescope interferometers. A laser is a light-emitting device based on electromagnetic radiation emissions using an optical amplification process. A laser differs from other light sources because it consistently emits light. Spatial coherence enables a laser to be tightly concentrated and allows for applications such as laser cutting and lithography. Spatial coherence also allows a laser beam to remain long distances narrow, allowing applications such as laser pointers. Lasers can also have a high time consistency that allows a very narrow spectrum to emit light, i.e. they can emit a single color of light. Temporary consistency can be used for light pulses that are as short as a femtosecond. Lasers are used in optical disk drives, laser printers and barcode scanners; fiber optical and open space optical communication; laser operation and skin treatments; cutting and welding materials; military and law enforcement devices for target marking and measuring range and speed; and laser lighting entertainment displays.

The book highlights the recent research and results in coherence and ultrashort pulse laser emission and related coherent phenomena of current interest. The book also deals with the generation of ultrashort pulse lasers and ultrashort pulse laser phenomena.

Table of Contents

Chapter 1

New Approaches in Frequency Conversion of Laser Radiation in Plasma Plumes

Rashid A. Ganeev

Institute of Ion-Plasma and Laser Technologies, Tashkent, Uzbekistan

ABSTRACT

Recent studies of plasma harmonic generation are reviewed. We show the advances of this technique for generation of coherent ultrashort pulses in the extreme ultraviolet range. Among the achievements highlighted in this review are the comparative experimental and theoretical studies of high-order harmonic generation in silver plasma, isolated subfemtosecond pulse generation in Mn plasma ablation, stable generation of high-order harmonics of femtosecond laser radiation from laser produced plasma plumes at 1 kHz pulse repetition rate, and high-order harmonic generation in fullerenes using fewand multi-cycle pulses of different wavelengths. We show that new developments of plasma harmonic studies can lead to the creation of an attractive field of laser ablation induced high-order harmonic generation spectroscopy.

Keywords: High-Order Harmonic Generation; Laser Plasma; Ultrashort Laser

1. INTRODUCTION

Coherent short wavelength radiation is of increasing importance for a broad variety of basic and applied research in various fields of physical, chemical, and life sciences. Among them, femtosecond time-resolved coherent diffractive imaging and photo-induced processes at surfaces and nanoparticles, as well as lithography, plasma diagnostics, and materials processing and diagnostics are of foremost interest. The high-order harmonic generation (HHG) from femtosecond laser pulses allows producing coherent radiation in the extreme ultraviolet (XUV) spectral range. Table-top lasers render these processes possible with the prospect of wide-spread scientific applications. Presently, predominantly few gases are employed as target media for HHG. Other efforts rely on the creation of harmonics at solid surfaces, either by a coherent wake field excitation or, for petawatt class lasers, on a relativistically moving electron gas acting as a plasma mirror. So far, however, only low conversion efficiencies have been obtained, despite the enormous efforts. Other methods include XUV free-electron lasers and X-ray lasers.

To promote the use of XUV radiation it seems therefore appropriate to advance laboratory scale sources to a higher application level. Many interesting experiments can be performed by HHG based on laboratory scale femtosecond lasers. These sources may cover the spectral range between 10 and 100 eV photon energy of harmonics, and with few-cycles laser systems even up to several 100 eV. For practical applications of high-order harmonic sources higher conversion efficiency and thus an increase in the photon flux and also of the maximum photon energy of the harmonic radiation would be beneficial. HHG itself can be used as a spectroscopic tool for analysis of the optical, nonlinear optical and structural properties of the emitters of harmonic generation presently comprising on a few noble gases. The generation of high-order harmonics in laser-produced plasmas from various solid-state targets, being for this purpose a relatively new and largely unexplored medium, promises to yield these advances.

Most interestingly, recent studies have shown that enhanced high-order harmonics can be generated also from the ablated nanoparticles, which opens the prospects for applications of local field enhancement, broad plasmonic resonances and a more efficient recombination processes for plasma HHG. As a highly interesting perspective an increase of the harmonic output by quasi phase matching in specially prepared plasmas may be considered. For the plasma a different and more flexible technique than used in neutral gases can be applied. The plasma may be spatially modified using a long pulse co-propagating with the fundamental driving pulse, and conditions might be found where quasi phase matching is possible over a long distance in the plasma, while the constructive and destructive interference of harmonic waves in such plasmas containing different emitters can provide a new knowledge about the phase-related characteristics of this process. Thus the above approach can be useful for producing an efficient source of short-wavelength ultra-short pulses for various applications and studies of the properties of harmonic emitters. The laser ablation induced high-order harmonic generation spectroscopy is a new method for the studies of material science and can be considered as one of the most important applications of HHG.

In this review, we discuss the realization of new ideas emerged during last few years, which further improved the HHG efficiency through harmonic generation in specially prepared plasmas and allowed the spectral and structural studies of matter through the plasma harmonic spectroscopy. We also present the current status of plasma HHG studies, and show new trends and perspectives in the developments of this filed.

The structure of the review is as follows. In Section 2, the comparative experimental and theoretical studies of high-order harmonic generation in silver plasma are discussed. Isolated sub-femtosecond pulse generation in the XUV range using Mn plasma ablation is analyzed in Section 3. Section 4 is devoted to the methods of stable generation of the high-order harmonics of femtosecond laser radiation from laser produced plasma plumes at 1 kHz pulse repetition rate. HHG in fullerenes using fewand multi-cycle pulses of different wavelengths is discussed in Section 5. Short summary remarks presented in Section 6 give some clues on future developments and perspectives of this interesting field of optics.

2. STUDIES OF HHG CUT-OFF IN SILVER IRRADIATED BY FEMTOSECOND PULSES

In this Section, we present the results of theoretical and experimental study of HHG in the ensemble of silver particles irradiated by intense femtosecond pulses of Ti:sapphire laser. It is shown that the photoemission spectra exhibit unusual behavior when the laser field strength approaches near-atomic values. In subatomic field strength the cut-off frequency increases linearly with laser pulse intensity. However, when the field strength approaches near-atomic region firstly cut-off frequency slows down and then saturates. To give new interpretation of such kind of photoemission spectrum behavior the light-atom interaction theory based on the use of eigenfunctions of boundary value problem for "the atom in an external field" instead of the traditional basis of the "free atom" eigenfunctions has been proposed.

In spite of the twenty-year history, the effect of the HHG is still under a great interest of both experimentalists and theoreticians. The origin of this interest is manifold. From a practical point of view, the HHG is one of the effective mechanisms for producing a coherent emission in a broad range of electromagnetic wave spectrum. The presence of plateau region in the harmonic amplitude distribution in XUV region affords grounds for development of sub-femtosecond pulse formation methods. As a result, the new frontiers are opened up in science by extending the nonlinear optics and time-resolved spectroscopy to the XUV range [1] and pushing ultrafast science to the attosecond domain, enabling the XUV spectroscopy and imaging of molecular orbitals [2], surface dynamics [3], and electron motion. The HHG is the reliable route to produce attosecond light pulses [4,5] and is therefore fundamental to attosecond science [6].

At present days, the efficiency of conversion to highorder harmonics is really too small to consider this emission as a real coherent XUV source for biology, plasma diagnostics, medicine, microscopy, photolithography, etc. Hence, the search for ways of increasing the cut-off frequency and the HHG efficiency in the XUV spectral range is still among the most topical problems of nonlinear optics.

The plasma HHG was observed with the large number of periodic table elements having usually small and middle atomic numbers [7-12]. As a rule, the interaction medium is a laser plasma that is prepared by irradiation of metal surface by picosecond pulses. The maximum harmonic order, or the cut-off frequency (CF), obtained in plasma media to date varies from the sixties to seventies harmonics of fundamental frequency [12] of Ti:sapphire laser. The highest-order harmonics (the 101st harmonic, λ = 7.9 nm) have been obtained in manganese plasmas [9]. The efficiency of conversion in the plateau region amounts to ~10−5 [10]. This value depends on both atomic energy level diagram and laser pulse parameters (intensity, energy and duration, carrier-envelope phase). The CF movement into the sub-nanometer region promises new possibilities in creating XUV coherent sources. Hence, the study of the CF dependence on the irradiated media properties and external laser pulse parameters is the problem of significant scientific and technical interest. Recently, the CF values corresponding to sixties and above harmonics have been obtained in experiments with silver plasma [10,11] and this stimulated to choose this medium for numerical simulations and compare the obtained results with the experimental data.

The traditional interpretation of the harmonic generation is usually based on the model of three-stage process that comprises the ionization of an atom, the electron acceleration in the electromagnetic field, and its subsequent collision with ion. This process is periodically repeated every half cycle of the electromagnetic wave. The evolution of the ionized electron is usually described with the help of Volkov wave functions or classical mechanics equations. It should be noted, that the use of the Volkov wave functions is due to the temporal evolution of the continuous spectrum electron wave functions in the presence of the laser field. At the same time, the spatial profile of the discrete spectrum wave functions is also changed in the presence of the external electromagnetic field. However, in the limit of zero strength fields the Volkov wave functions are transformed to the free electron wave functions, while the discrete spectrum wave functions became the atomic eigenfunctions. Hence in the weak or zero fields the eigenfunctions of discrete and continuous spectra are not mutually orthogonal and they do not compose the complete basis of the orthogonal functions. This is the most principle inconsistence of the above methods.

In those studies, the analysis was focused on the single silver atom photoemission spectra [13]. The most unusual feature of the single atom photoemission spectra is the CF saturation effect. The CF saturation means that the maximum observed harmonic order ceases to be intensity dependent at $I \geq I_{sat}$. For the first time, the effect of the CF saturation in silver plasma was experimentally observed in [10]. The maximal order of experimentally observed harmonic is equal to 71. On the other hand, the results of computer simulations performed for the single silver atom [14] have shown that the CF is saturated in the region of over atomic field strength, but at near atomic field strength the maximal order of generated harmonics is much higher than that observed experimentally. The main purpose of the presented study is to explain the difference between the experimental and theoretical results and to provide the interpretation of the main mechanisms lying in the origin of the CF saturation effect.

To study the high-order nonlinearities through the HHG, the silver target was placed inside the vacuum chamber to ignite the plume by laser ablation of this sample. Silver has previously proven to be most efficient medium for production of the plasma plumes where the highest HHG efficiency (8 × 10−6) in the XUV range was observed [15].

The pump source was a commercial, chirped pulse amplification laser system, whose output was further amplified using a homemade three-pass amplifier operated at a 10 Hz pulse repetition rate. A heating pulse was split from the amplified laser beam by a beam splitter before a pulse compressor (Figure 1). The heating pulse duration was 210 ps. A spherical lens focused the heating pulse on the target placed in the vacuum chamber. The area of ablation was equal to 0.5 mm2. The heating picosecond pulse intensity at the target surface was varied in the range of Ipp = (0.5 – 5) × 1010 W·cm−2 by changing the energy of heating pulse. A driving femtosecond pulse at a centre wavelength of 792 nm had the energy of 12 mJ and pulse duration of 150 fs after propagation of the compressor stage. After the proper delay with regard to the heating pulse irradiation (20 - 80 ns), the driving pulse was focused by a spherical lens (f/10) on the ablation plume from the orthogonal direction. The maximum intensity of the driving pulse at the focal spot could reach 6 × 1016 W·cm−2. The position of laser focus was adjusted to be either before or after the laser plume to exclude the influence of considerable ionization of the silver neutrals and ions in laser plume and to maximize the harmonic yield. The intensity of the driving femtosecond pulse (Ifp) at the plasma area was adjusted between 1×1014 and 1×1015 W·cm−2. The distance between the HHG interaction area and target surface was varied in the range of 100 - 200 mm. The spectrum of generated highorder harmonics was analyzed by a grazing incidence XUV spectrometer with a gold-coated flat-field grating. The XUV spectrum was detected using a microchannel plate (MCP), and the optical output from the phosphor screen was recorded using a charge-coupled device (CCD) camera.

Figure 2(a) shows the image of harmonic spectrum from Ag plasma at the conditions when the femtosecond pulse was focused after the plasma plume. The harmonics up to the 55th order were routinely observed at these conditions, when the intensity of femtosecond pulse at the plasma area was Ifs = 2.5 ´ 1014 W·cm−2. Note that divergences of harmonics at the beginning of plateau-like distribution of harmonics and at the cut-off region were approximately equal to each other. The increase of laser intensity at the plasma area led to increase of harmonic cut-off (Hc) up to the 61st order. The divergence of harmonics gradually increased with the growth of harmonic order and the divergences of the lowest (17th - 25th) were approximately two to three times less compared with the highest (45th - 57th) harmonics.

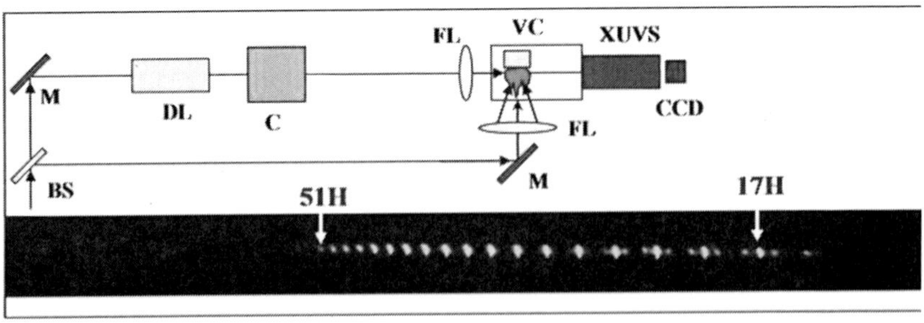

Figure 1. Experimental setup for HHG in silver plasma. BS, beam splitter; M, mirrors; DL, delay line; C, compressor, FL, focusing lenses; VC, vacuum chamber; XUVS, extreme ultraviolet spectrometer; CCD, charge coupled device. Inset: Typical image of harmonic distribution from 17th to 51st orders [13].

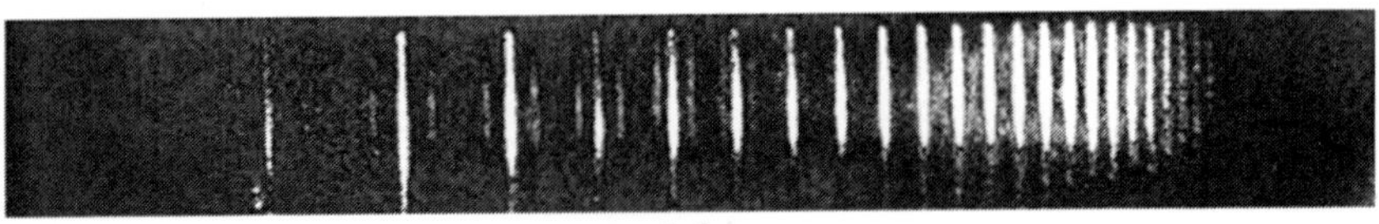

(a)

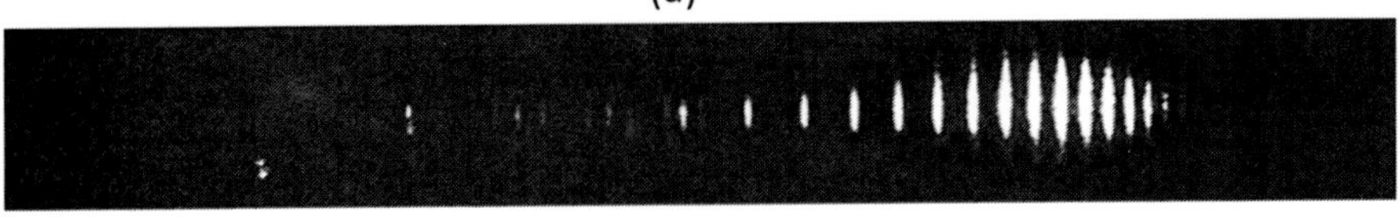

(b)

Figure 2. (a) Harmonic spectra from silver plasma in the case of focusing behind the plasma area; (b) Spectral image of harmonics from silver plasma in the case of aperturing the input femtosecond radiation. Numbers show the harmonic orders [13].

At these experimental conditions (focusing inside the plasma plume, intensity of the driving pulse at the plasma area Ifs = 4 ´ 1014 W·cm−2, delay between the pulses of 30 ns) the divergence of driving radiation was changed by introducing the aperture in front of the input window of vacuum chamber through which femtosecond radiation propagated toward the plasma plume. A decrease of beam sizes led to considerable deformation of observed spectra, when one can find the conditions of variations of the divergence of harmonics in different ranges of harmonic spectra, as well as considerable re-distribution of harmonic efficiency along the plateau region. When the aperture was decreased down to 4 mm (and correspondingly decreased the energy of propagated radiation from 12 mJ to 4 mJ) a remarkable pattern of harmonic spectra appeared showing the five to ten times enhancement of harmonic intensity along the axis of laser propagation for higher harmonics compared with lower ones (Figure 2(b)). The on-axis intensities of 11th - 19th harmonics were considerably less than those in the range of 25th - 47th orders.

The Hc(Ifp) dependence became saturated at Ifp » 4 × 1014 W·cm−2 (see Figure 3" target="_self"> Figure 3 showing the dependence between the maximum observed harmonic order and the femtosecond pulse intensity). This intensity is well above the one at which the neutral Ag atoms are ionized. So it may be an additional indication of the crucial role of singly charged ions in generation of high-order harmonics.

The novelties of these experiments are 1) the observation of the efficient harmonic generation for the high orders, while keeping the low-order ones at smaller conversion efficiency; and 2) the observations of saturation of the HHG yield above some level of laser intensity (4 × 1014 W·cm−2). The aperturing of laser beam led to better phase matching conditions for higher orders compared with lowest harmonics (Figure 2(b)). The intensity of those high-order harmonics became strong enough to allow the observations of second-order diffraction patterns from the grating (see the lobes from the left and right sides of lowest harmonics on Figure 2(a)).

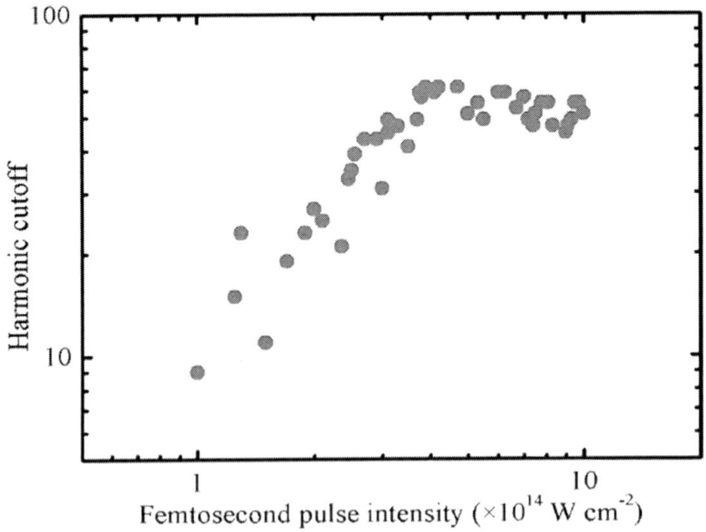

Figure 3. Harmonic cut-off from silver plasma as a function of laser intensity. The saturation of Hc(Ifp) dependence was started at ~4 × 1014 W·cm−2 [13].

The violation of the phase-matching conditions at high intensities of driving radiation is caused by appearance of the additional free electrons possessing positive dispersion, which does not match with the relatively low dispersion of atoms and ions. The difference with the case of analogous observations in the gaseous media is related with presence of the free electrons at the initial state of plasma media. The ablation at appropriate conditions can create the formation of "optimal" plasma, which, whilst containing free electrons, does not prevent the violation of phase matching. Only additional free electrons appearing during further ionization of singly charged ions of silver (second ionization potential of silver is 21 eV, which corresponds to the breakdown saturation intensity of laser radiation of ~3 × 1014 W·cm−2).

Silver is one of most attractive targets for plasma formation and HHG. Conversion efficiency reported for this plasma medium for the harmonics in the range of photon energies of 50 - 80 eV (8 × 10−6, [10]) is the highest among other plasma plumes. Moreover, the value of conversion efficiency is the same, as those reported in the case of gas HHG studies. Assuming the stability amendments in the harmonic generation from this plasma (of the order of 1 million shots at 1 kHz pulse repetition rate [16]) one can anticipate a robust application of this target for various applications of plasma HHG spectroscopy.

The process of a light-atom interaction becomes fundamentally the non-resonant process, when the laser pulse temporal width approaches to a few cycles of optical oscillations. Indeed, in this case the energy of such pulse is comparable or exceeds the energy level difference between the free atom states. Hence, the well developed theories of the resonant oneor multi-photon interactions cease to be valid in this case. Moreover, the binding energy of atomic electrons tends to zero when the laser field strength exceeds the intra-atomic field strength, because the electron moves mainly in the spatial regions, which are far from its parent nuclei. It is evident that the basis of a free atom eigenstates could not serve as a useful basis for the wave function expansion when the laser field strength approaches to the intra-atomic values. At the same time, as it has been shown in [17], there are direct mathematical relationships between the eigensolutions of the "free atom" boundary value problem and boundary value problem for "the atom in an external field".

The numerical calculations yield a series of harmonic emission spectra from silver particles. At small intensities, in subatomic region the spectrum consists of the sequence of the odd harmonics of incident field carrier frequency (Figure 4(a)). The amplitudes of harmonics sharply drop with the harmonic number. The origin of this emission is the light-induced (and, hence, time dependent) anisotropy of a spatial profile of an electron wave function: the electromagnetic wave induces the radial electron motion in the intra-atomic potential field and, hence,

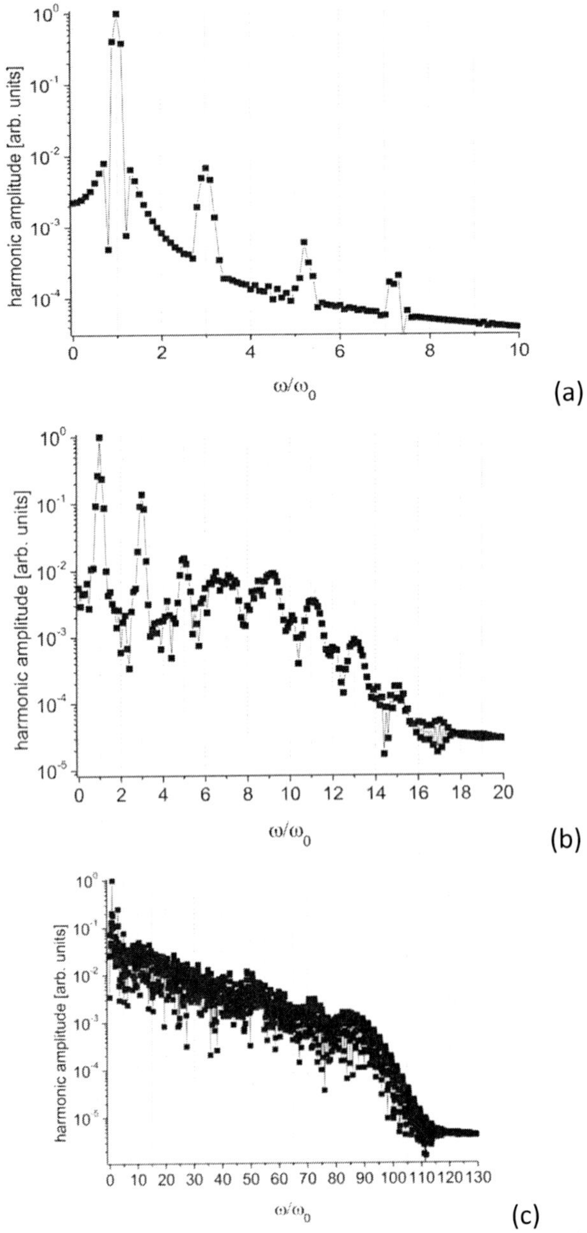

Figure 4. The photoemission spectra generated in silver at the laser pulse intensities of (a) 5.94 × 1010; (b) 1.83×1012, and (c) 1.88 × 1013 W·cm−2 [13].

produces some external work. This motion will inevitably result in the appearance of harmonics. At a given laser field strength this process could be due to elastic light-atom interaction, because after the termination of the laser pulse action the atomic electron can be again found in the ground state.

When the laser field strength approaches 3 × 1011 W·cm−2, the shape of the response spectrum is changed. At higher intensity, there appear tendencies for growing of its width and formation of plateau (Figure 4(b)). Further increase in the amplitude of the external field strength allows to observe well-defined plateau with definite CF (see Figure 4(c) for I = 1.88 × 1013 W·cm−2).

The results of computer calculations of the CF as a function of the laser pulse intensity is presented in Figure 5(a). The solid line in this figure shows the approximated analytical dependence obtained for a silver atom in the frames of the approximation discussed in [13]. In the region of the laser pulse intensity of 1011 < I < 1012 W·cm−2 the results of computer calculations are in good agreement with the proposed approximation. Under further increase of the laser pulse intensity the calculated curve raises faster than it is predicted by the proposed approximation. In this region of pulse intensity the population of other discrete excited states becomes comparable and even exceeds the population of $5p$ state of Ag due to the recombination process. Hence, the new channels of photoemission are revealed.

At laser pulse intensity I > 1014 W·cm−2 the CF is saturated, i.e., it ceases to be intensity dependent. The reason of such behavior is quite obvious and it is associated with the fact that the probability of an electron ionization approaches to unity in this region of a pulse intensity. It means that the atomic electron is almost entirely localized in the continuum spectrum states and it does not collide with its parent ion. At the same time, as far as the time profile of a laser pulse has relatively soft slope in the front of pulse, hence, the harmonics are effectively generated here. So, the results of computer modeling show that the most probable reason of the CF saturation is the total ionization of the irradiated atom.

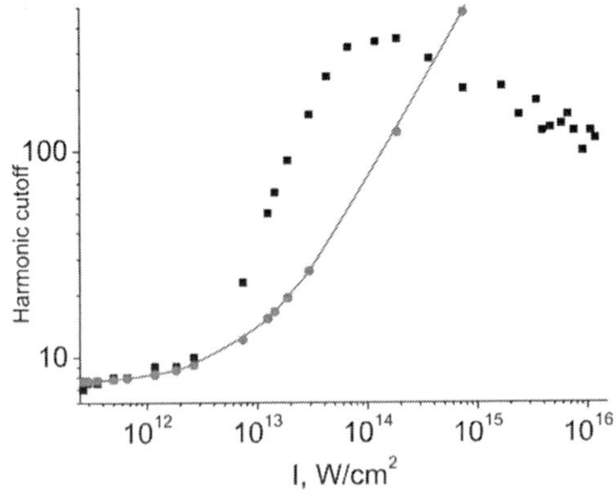

(a)

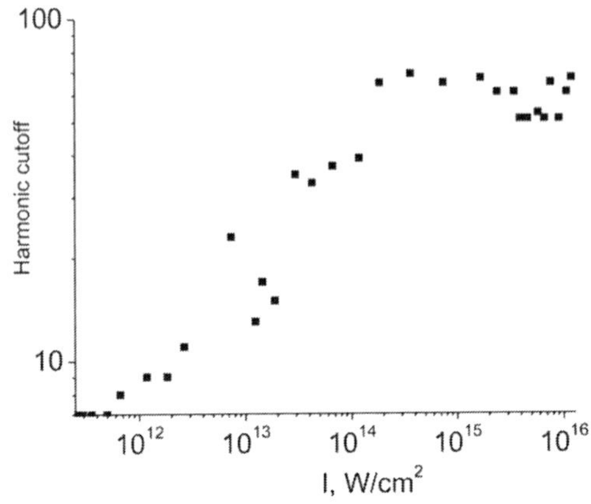

(b)

Figure 5. (a) Cut-off frequency as a function of the laser pulse intensity. Squares are the result of numerical calculations, solid line with circles is the dependence given by the approximation discussed in [13]; (b) Cut-off frequency of an ensemble of atoms as a function of a laser pulse intensity [13].

One can see from Figure 5(a) that the calculated laser pulse intensity Icalc » 1014 W·cm−2 at which the CF is saturated is close to the experimentally measured value (Iexp = 4 × 1014 W·cm−2). However, the calculated maximal order of harmonics exceeds the experimentally measured value for more than four times. In computer simulations we deal with a single silver atom. Hence it is natural to assume that the reason of this difference can be due to the frequency dependent dispersion appropriate to the spatially distributed ensembles of atoms.

The dispersion properties of a spatially distributed ensemble of atoms could result in drastic transformation of an atomic ensemble emission spectra in comparison with the single atom one. Figure 5(b) shows the integral dependence of the CF as a function of laser pulse intensity, which is calculated for spatially distributed ensemble of atoms. It was taken into account in these calculations that the ionization probability depends on the peak pulse intensity. It can be seen that the curve in Figure 5(b) fits much closer to the experimentally measured dependence.

Summarizing this section, the saturation of the CF of the photoemission spectra generated in a silver by a femtosecond laser pulses has been demonstrated both experimentally and theoretically. In experiments with 155 fs pulse of Ti:sapphire laser the highest experimentally measured harmonic was 61 and the saturation intensity of Iexp = 4 × 1014 W·cm−2. The theoretical analysis is based on the non-perturbative theory of nonlinear response, which enables to vary the laser field strength in a wide region from a subatomic to a over-atomic values. The results of computer simulations have shown that the CF is also saturated, but the maximal harmonic order exceeds three hundred in the case of single atom. This drastic difference between experimental and theoretical results is due to the dispersion effects, which play a decisive role in the spatially distributed ensemble of ionized atoms. At given experimental conditions the phase mismatch between the driving laser pulse and the harmonic emission modifies significantly the frequency spectra of an atomic ensemble in comparison with the single atom response. The amplitudes of high-order harmonics are significantly suppressed. The simplest way to avoid this modification is to decrease the atomic density. However, this way is in contradiction with the perspectives of practical applications of the high-order harmonic emission.

3. ISOLATED SUB-FEMTOSECOND XUV PULSE GENERATION IN MN PLASMA

In this section, we discuss HHG from transition metal plasmas. These are very promising targets in view of the giant resonances found in the photoionization cross sections. For example, the Mn+ cross section is ~40 Mb at 50 eV photon energy, whereas rare gas atoms have cross sections between 1 and 8 Mb at this photon energy. Photorecombination-the third step in the recollision model-is the inverse process of photoionisation and therefore HHG and photoionization must exhibit the same resonances. This has been confirmed not only by previous resonance-induced experiments with laser-produced transition metal plasmas but also in a recent study of HHG from xenon gas [18].

Resonance-induced enhancement of a single harmonic of the laser radiation allowed considerable improvement of harmonic efficiency in some specific XUV spectral ranges related with high oscillator strengths of ionic states of metals. This was confirmed in multiple studies following the initial observation of this phenomenon in indium plasma [19]. In particular, strong enhancement of a single harmonic was reported in Cr [20] and Mn [9] plasmas. The Mn plasma is of special interest since it showed the highest harmonic cut-off energy observed in plasma plumes (101st harmonic of Ti:sapphire laser [9]). In previous studies, multi-cycle (30 fs [9] and 140 fs [21]) laser pulses were employed and the generation of all harmonics in the plateau was observed together with a strongly enhanced harmonics in the vicinity of 50 eV.

Recent progress in the generation of few-cycle pulses allowed the observation of various new effects including the realization of isolated attosecond pulse generation in gas media [22-24]. In this connection it is interesting to analyze resonance-induced processes observed in an ablation plume using the shortest available laser pulses. Below, we present the analysis of the experiments on resonance enhancement in manganese plasmas using 3.5 fs pulses [25]. The most interesting feature observed in those experiments was a suppression of almost all neighboring harmonics in the vicinity of a resonantly enhanced single harmonic at the photon energy of ~50 eV.

The experimental arrangements were analogous to those presented in previous section of this review, excluding the use of few-cycle pulses for the experiments under consideration. The harmonic spectrum in the case of propagation of the 3.5 fs pulses through the manganese plasma was strikingly different compared with other plasma samples (for example Ag plasma) analyzed in separate experiments. While all other samples studied showed a relatively featureless harmonic spectra with extended cut-off (Figure 6, upper curve showing spectrum of the harmonics generating in the silver plasma), the Mn plasma allowed generation of a strong single harmonic substantially enhanced compared with neighboring ones (Figure 6, bottom curve).

As it was already mentioned, the harmonic spectra from manganese plasmas for 30 fs and 140 fs pulses also showed enhanced harmonics around 50 eV. The assumption of the resonance nature of the enhancement of harmonics of the ~800 nm radiation of Ti:sapphire lasers in this spectral region is supported by the presence of a strong giant resonance in the vicinity of 50 eV confirmed by experimental [26] and theoretical [27] studies. The enhancement of a single harmonic can be attributed to the broadband resonances of the ions of few metals, such as V, In, Cd, Cr, Cd, and Mn. These “giant” resonances have been experimentally confirmed in the literature [26,28,29] and discussed recently in a few theoretical studies [7,30-32].

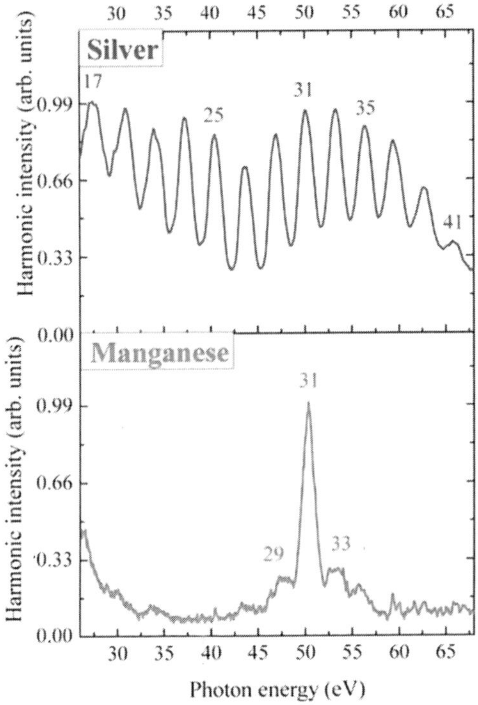

Figure 6. Harmonic spectra from the silver plasma (upper curve) and manganese plasma (bottom curve) [25].

However, in previous studies using multi-cycle pulses, the intensity of enhanced harmonics was only a few times higher compared with those neighboring harmonic orders. The same features were reproduced in the reviewed studies using 40 fs pulses from another Ti:sapphire laser at similar intensity inside the laser plasma (4 × 1014 W·cm−2). The raw image of the harmonic spectrum presented in Figure 7(a) shows several enhanced harmonics starting from the 31st order followed by an extended second plateau. The extension of the harmonic cut-off exceeding the 71st order is attributed to the involvement of doubly charged Mn ions as the sources of HHG. This feature of Mn plasma harmonics has already been reported earlier [9]. Here also presented a typical image of a Mn harmonic spectrum in the case of 3.5 fs pulses

(Figure 7(b)). No second plateau, which was seen in the case of multi-cycle (40 fs) pulses, is observed for the few-cycle pulse. Most striking was the observation of a single, very strong, broadband (2.5 eV)

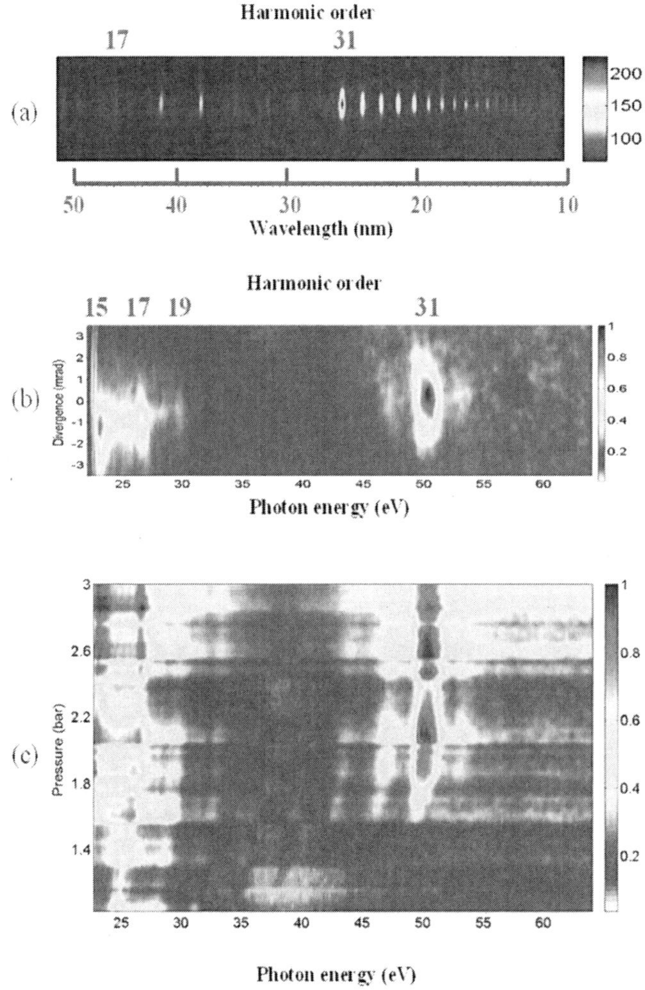

Figure 7. Raw images of harmonic spectra from manganese plasma in the case of (a) 40 fs and (b) 3.5 fs driving pulses obtained at the same intensity; (c) Raw images of harmonic spectra from Mn plasma at different pressures of neon in the hollow fiber obtained at the same energy of driving laser pulses [25].

31st harmonic. Only two weak neighboring harmonics (around the strong emission) are seen in the 30 - 65 eV spectral range. The ratio between the intensities of the enhanced harmonic to the weak neighboring harmonics exceeded one order of magnitude. One can note that, at a lower intensity of the femtosecond pulse (<2 × 1014 W cm−2), this strong harmonic disappeared when using both multiand few-cycle pulses.

The distinctive structure of the harmonic spectra, both for 40 fs and 3.5 fs pulses, clearly points out the involvement of Mn resonances centered around 50 - 51 eV. The same can be said about the photoionization or photoabsorption characteristics of Mn+ plasma, which are due to the "giant" 3p → 3d resonance [28]. The laser polarization dependence of this emission was analyzed and it was found that the 50 eV radiation abruptly decreases with the change of the polarization state of the femtosecond driving pulses from linear to elliptical, which is a clear signature of the emission being due to high-order harmonic generation.

To analyze the effect of the spectro-temporal characteristics of the femtosecond radiation on the harmonic yield, the pressure of neon in the hollow fiber of second compressor of laser was varied, thus changing the duration of driving pulse. The spectral and intensity variations of manganese harmonic spectra in the range of 22 - 62 eV as the functions of neon pressure in the hollow fiber are shown in Figure 7(c). One can clearly see that, with change of pressure (from 1 to 2.3 bar), the single 31st harmonic intensity varies from almost zero to its maximum high value. A blue shift of the harmonics is also evident. Further increase of neon pressure up to 3 bar, at which the experiments with the 3.5 fs pulses were carried out, did not change the harmonic distribution.

The experiments described above were carried out without carrier envelope phase (CEP) stabilization (that is, for random CEP values). The HHG experiments with Mn plasma using 3.5 fs pulses were also performed with stabilized CEP (e.g. at phases values of $\phi = 0$ and $\pi/2$) and no considerable differences were found in that case (Figure 8), though some variation of harmonic distribution was observed for the lower order harmonics (compare the middle and bottom curves of Figure 8). The spectral shapes of the 31st harmonic emission were quite similar for these two fixed values of CEP, while a considerable difference in harmonic spectra was maintained when comparing to longer pulse duration and lower intensity of the driving pulses. Figure 8 shows HHG spectra for 25 fs pulses (upper panel) and 3.5 fs pulses (middle and bottom panels) of the same energy. One can clearly see the absence of harmonic extension and resonance-induced HHG in the case of low-intensity, 25 fs pulses.

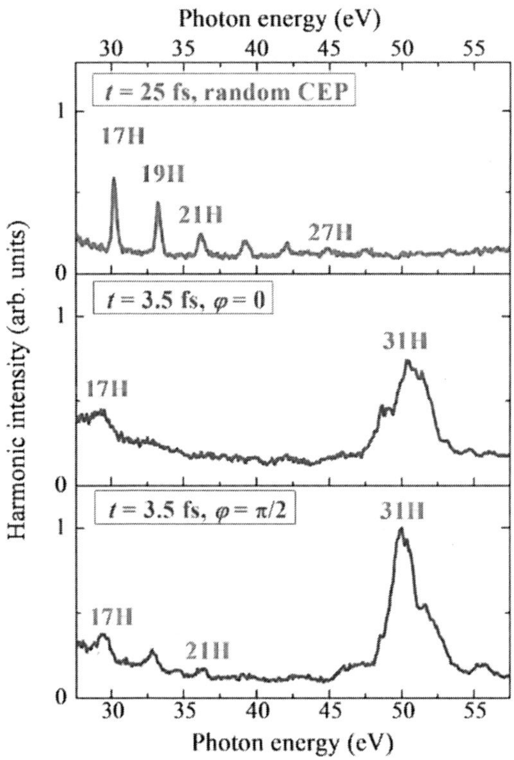

Figure 8. Experimental harmonic spectra generated from manganese plasma in the case of the absence of gas in the hollow fiber compressor (t = 25 fs) and random CEP (upper panel), and at 3 bar pressure (t = 3.5 fs) at fixed CEP (ϕ = 0, middle panel; ϕ = π/2, bottom panel). ϕ denotes the carrier envelope phase [25].

The fact that a strong CEP dependence of the plasmaharmonic spectra in the case of 3.5 fs pulses was not observed could also be attributed to the presence of a significant amount of free electrons in the manganese plasma, which might diminish the difference between the HHG spectra recorded for different values of CEP. The same can be said about other HHG experiments using silver and brass plasmas, which did not show significant differences in harmonic spectra when comparing fewcycle pulses with fixed and random CEP. In the meantime, comparative studies with gas media under similar experimental conditions were carried out and found a characteristic dependence of the HHG spectra on the CEP, which were commonly observed at such conditions. Thus the absence of the influence of the CEP on the harmonic pattern generated by few-cycle pulses from the ablation plumes appears to be a specific feature of plasma HHG.

Summarizing this section, the observation of harmonic spectra from manganese plasmas driven by 3.5 fs pulses, which are dominated by a single enhanced 31st harmonic at around 50 eV, was discussed. The spectro-temporal experiments reveal the influence of the drive pulse duration on this process. Analysis of the harmonic energy tuning upon the variation of driving laser pulse duration controlled by change of gas pressure in the hollow fiber compression system show stabilization of the enhanced harmonic's energy and wavelength, which can be explained with resonance-induced enhancement at 51 eV. The modeling of this process suggested that the emission could constitute an isolated sub-femtosecond pulse. The observed weak CEP dependence might reduce the requirements for CEP stabilization of the laser. The isolation of a single harmonic without any filtering could also be useful for various applications where the ultrashort XUV pulses are required, without the losses induced by spectral dispersion or filtering.

4. STABLE GENERATION OF HIGH-ORDER HARMONICS OF FEMTOSECOND LASER RADIATION FROM LASER PRODUCED PLASMA PLUMES AT 1 KHZ PULSE REPETITION RATE

Almost all HHG studies from weakly ionized plasmas produced during laser ablation of various solid targets were carried out using the 10 Hz pulse repetition rate lasers [11]. Up to now only few studies of HHG from plasmas have been carried out on static targets using 1 kHz class lasers [33-35]. The ablation process at 1 kHz pulse repetition rate causes a considerable change of the surface properties of the target due to the melting, which deteriorates the plasma plume conditions during laser ablation. Surface heating and melting of a static target result in an unstable harmonic signal so that movement of the target surface is required to maintain a reasonable stability. The demand in finding the optimal way for improving the plasma harmonic stability at 1 kHz pulse repetition rate is high due to recent observations of the advanced properties of plasma harmonics over gas harmonics [36-38]. In particular, in [36], the plasma HHG conversion efficiency was measured one order of magnitude stronger compared with gas HHG efficiency. Analogous features were reported in [38].

The obstacle of all plasma harmonic experiments during earlier studies was an insufficient stability of plasma parameters (density, ion and free electron concentrations, excitation conditions, etc.), which led to an instability of harmonic yield and fast decay of harmonic efficiency during irradiation of the same spot of ablating target. As it was mentioned, most of those early studies were performed using a 10 Hz class lasers. At this relatively low pulse repetition rate, the stability of harmonics deteriorated after a few hundred shots on the same spot of the surface and even quicker for powder-like materials (fullerenes, nanotubes, metal nanoparticles, various organic and non-organic powders). One can note that laser ablation of the latter samples can be considered as an important tool for their structural studies using XUV nonlinear spectroscopy. The application of soft ablation allows the use of the same target for a much longer period than in the case of earlier studies of over-excited targets during laser ablation. Thus a search of a robust, easy-to-apply method for improving the harmonic stability in the case of plasma HHG could considerably advance a search of the fundamental (structural, orientational, etc.) properties of organic and inorganic atoms and molecules.

The earlier used approaches of a rotating disc geometry [39,40] for the movement of targets during ablation are not suitable since the distance between driving femtosecond beam and target surface should be maintained minimal (of order of 100 mm), while the Rayleigh length of the driving beam is maintained at the range of a few mm. Below we discuss a new method using a motorized rotating rod specifically prepared for the HHG from plasma plumes using high pulse repetition rate lasers, and demonstrate that this target significantly improves the stability of high-order harmonics [16].

Those studies were performed using two laser pulses: one to produce the plasma plume and the second to drive the HHG within it. The first (heating) pulse was created by splitting off a portion (200 μJ) of the uncompressed 8 ps laser pulse from a 1 kHz Ti: sapphire chirped pulse amplification laser. The remaining pulse was compressed in a prism compressor and then further compressed using a hollow core fibre and chirped mirrors, resulting in 250 μJ, 3.5 fs pulses. The driving pulse was delayed with respect to the heating pulse by 40 ns to give the plasma the time to expand away from the target surface to allow the driving pulse to pass through the plasma without being clipped by the target.

Target rotation and movement apparatus consisted of three linear stages driven by stepper motors along three axes. The target was attached to an axis of the fourth motor, which provided rotation with a variable speed (from a few rotations per minute (rpm) up to 300 rpm). Rotating the target was sufficient to achieve stable harmonic radiation and an additional vertical movement was not required, though this capability might be useful for future plasma HHG experiments. As the setup requires the target to be positioned very close to the driving beam, it was of paramount importance that the target was carefully aligned to the axis of rotation. Any movement of the target surface due to eccentricity in the radial direction from the driving beam axis would result in an oscillation of the harmonic signal due to variation of the plasma density seen by the driving beam or, in the extreme case, clipping of the laser beam.

The target (cylindrical rode with diameter of 10 mm and length of 30 mm) was positioned as shown in Figure 9, with the driving pulse propagating ~100 μm above the target surface. The picosecond heating pulse was focused onto the surface of the rotating target. In order to efficiently produce high-order harmonics the plasma must be weakly ionized. To achieve this the target was positioned slightly in front of the focus of the heating pulse using a 50 cm focusing lens, leading to an on target intensity of ~ 1×1010 W·cm−2. This also had the benefit of increasing the size of the plasma produced from ablating a larger area. The size of the focus at the target surface was measured to be ≈500 μm. The delayed driving pulse was focused through the plasma using a 40 cm spherical mirror. The HHG radiation was analyzed by an XUV spectrometer consisting of a flat-field grating and an imaging microchannel plate detector with phosphor screen imaged onto a CCD camera.

Figure 10 shows that there is a drastic change in the harmonic signal (integrated over the spectral range of 40 - 80 nm) when the rotation of the aluminum target is stopped. There is a sharp intensity decrease of more than one order of magnitude over only one thousand shots (or just after one second of ablation using 1 kHz laser). The benefits of the rotating rod are clearly shown in Figure 11 where stable harmonic generation was achieved from the plasma produced on an aluminum target for over 1 million laser shots. Stable harmonics were achieved in a broad range of the speeds of rotation. The target rotational speed and the size of the ablation focus imply that the same area of target was undoubtedly used repeatedly for consecutive rotations over the 20 minute duration of experiments.

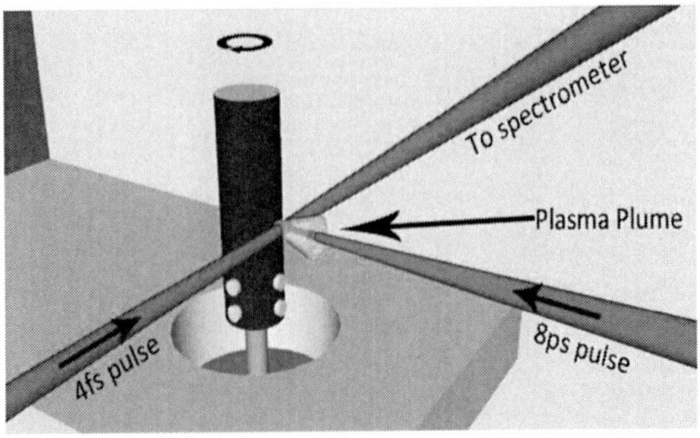

Figure 9. Schematic of rotating target and HHG configuration [16].

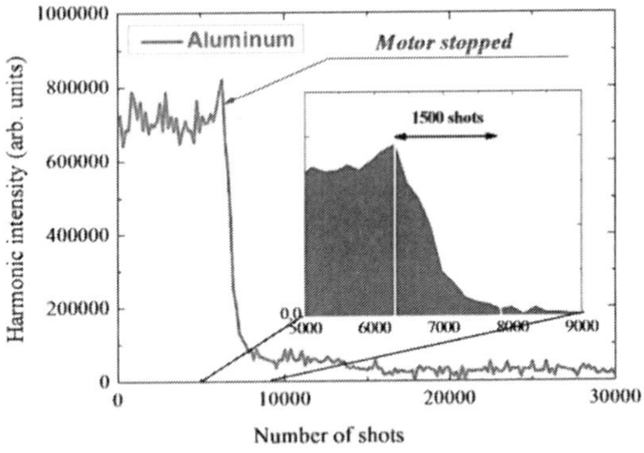

Figure 10. Decay of harmonics from aluminum plasma after stopping the rotation of the motor. The harmonics were integrated over 40 - 80 nm spectral range [16].

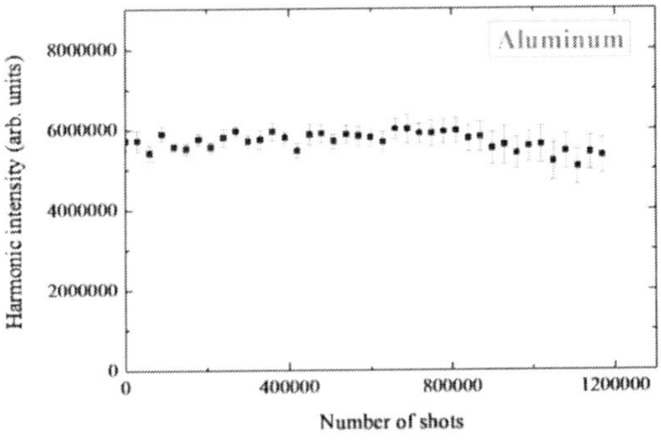

Figure 11. Stability of integrated harmonic signal from aluminum plasma in the case of rotating target [16].

It is possible that once the fixed surface is melted the force from a following laser shot and plasma creation could expel some of the liquid target from the ablation area, which would not cause the plasma to be emitted in a direction normal to the surface. These effects are considerably diminished once the target starts to rotate. During rotation, the previously ablated area cools down such that, during the next set of ablation on this spot, the plasma formation occurs at approximately same conditions. To prove that the ablated area cools down with rotation, the target was rotated at different speeds (from 10 to 300 rpm) and no difference in stability of harmonic yield was found. These observations point out the importance of the periodic change of the ablation zone. This also confirms a suggestion that the cooling of the ablation area leads to stable plasma generation.

Characteristics of plasma (density and ionization state) are the most important parameters to achieve and maintain stable HHG efficiency during an extended period of target illumination. The calculations [38] have shown that, in the case of carbon plasma, the concentration of particles in the area of femtosecond laser-plasma interaction at optimal delay between heating and driving pulses (~40 ns) is ~2 ´ 1018 cm−3. The solid surface was considered as the one unheated before the laser ablation. Indeed, after one round of rotation (e.g., after 0.2 - 2 s), the plasma disappears, the ablated spot cools down, and the next laser shot on the same spot can be considered as a shot on the almost "fresh" surface. Contrary, in the case of a stationary target, the following shots continue the heating of the same spot.

Summarizing, the novelty of this approach includes the observation of advanced properties of plasma HHG even at extremely small energies of the heating pulses. The efficiency of plasma HHG depends on the possibility to create "optimal" plasma. This can be done using both multi-mJ pulses, as was shown in previous studies [38], and few hundred mJ pulses, as it was demonstrated in the reviewed work [16] and recently published studies [36]. The important point here is the intensity and fluence of the heating pulse on the target surface. The application of a higher energy heating pulse could create the conditions of "optimal" plasma over a longer distance, which could (or could not) increase the harmonic yield depending on the phase relations between the driving and harmonic waves. In addition, it can also lead to the over-heating of the target at 1 kHz ablation. As it was already mentioned above, the rotating speed did not influence the stability of harmonics using 0.2 mJ heating pulses. The use of more energetic pulses at high repetition rate (i.e. of order of few mJ) may require additional optimization of the rotation target technique (for example, by periodic up and down dragging of the rotating target).

5. HIGH-ORDER HARMONIC GENERATION IN FULLERENES USING FEWAND MULTI-CYCLE PULSES OF DIFFERENT WAVELENGTHS

Finding new approaches to improve the efficiency of HHG of laser radiation towards the XUV range is an important goal of nonlinear optics and laser physics. Small sized nanostructures are an attractive approach, since they exhibit local-field-induced enhancement of the nonlinear optical response of the medium [41-43]. Another mechanism that can enhance harmonic efficiency of clusters is the increase of the recombination cross section of accelerated electron and parent particle in the final step of the three-step mechanism of HHG [44]. It was shown in [45] that laser-irradiated cluster-contained plasmas can emit low-order harmonics efficiently. The advantages of highly efficient harmonic generation in laser-produced plasmas containing nanoparticles have been analyzed in [43]. In this connection small-sized clusters, such as fullerenes, could show the advanced nonlinear optical properties Fullerenes are an attractive nonlinear medium for the HHG. Their sizes and broadband surface plasmon resonance (SPR) in the XUV range allowed the first demonstration of enhanced HHG near the SPR of C60 (λSPR » 60 nm, with 10 nm full width at half maximum) [46]. Theoretical studies of HHG from C60 using multi-cycle pulses include an extension of three-step model [47], analysis of the electron bound within a thin shell of a rigid spherical surface, with geometrical parameters similar to those of the C60 [48], and application of dynamical simulations [49]. Those studies reveal how HHG can be used to probe the electronic and molecular structure of C60. Theoretical investigation of such systems is hampered by the fact that the Hamiltonian of HHG is time dependent and the systems consist of many electrons. The investigation of the influence of the electrons on the resonant HHG can be performed by means of a multiconfigurational time-dependent Hartree-Fock (MCTDHF) approach, which has the accuracy of direct numerical solution of the Schrödinger equation and is not significantly more complicated than the ordinary time-dependent Hartree-Fock approach [7]. In particular, the computations could be based on the multiconfigurational time-dependent Hartree software packages [50]. In [7], simulations of resonant HHG were performed by means of a MCTDHF approach for three-dimensional fullerenelike systems. The influence of the SPR of C60 on the harmonic efficiency in the range of 60 nm (E = 20 eV) was analyzed and showed the role of resonant effects in the HHG enhancement.

The ionization saturation intensities of different charge states of C60 are higher compared to isolated atoms of similar ionization potential [51]. With this perspective, it is interesting to analyze HHG from fullerene molecules in the field of few-cycle laser pulses and compare these studies with those carried out using multi-cycle pulses. The motivation for the discussed studies was to analyze the conditions for efficient HHG from plasmas containing C60, using picosecond laser pulses to ablate the fullerene-containing target at high pulse repetition rate (1 kHz) and then few-cycle laser pulses (t = 3.5 fs) to generate the harmonics in the fullerene plasma [52]. We also analyze HHG in fullerenes using 1300 nm radiation and compare these studies with those using 780 nm multicycle pulses.

The experimental setup using a few-cycle pulses for fullerene harmonic generation was analogous to those presented in sections 3 and 4. Two types of targets were investigated: 1) C60 powder, which was glued onto a glass substrate or onto a rotating aluminum rod; and 2) bulk graphite for plasma harmonic generation. The maximum harmonic conversion efficiency was obtained by varying the distance between the femtosecond beam and the target, as well as the z-position of the plasma. The harmonic spectra from plasmas produced on the bulk graphite and fullerene powder glued onto the glass surface are presented in Figure 12(a) in the case of laser wavelength λ = 780 nm. Harmonics up to the 29th order were obtained from the fullerene plasma. The harmonic efficiency in the case of the graphite plasma was about five times higher compared with the case of fullerene plasmas, most likely due to the higher concentration of emitters in the former case, as has also been reported in [37].

The results of comparative studies of the HHG in fullerene plasma using few-cycle (3.5 fs) and multi-cycle (40 fs) pulses are shown in Figure 12(b). In the latter case, a 1 kHz repetition rate Ti:sapphire laser generating 4 mJ, 40 fs pulses was used. The HHG conversion efficiency for the 40 fs pulses at the beginning of plateau range in the case of the plasma plume containing fullerenes was estimated to be ~5 × 10−6 using a comparison with the HHG conversion efficiency in a silver plasma, which has previously been reported at similar experimental conditions to be 1 × 10−5 [43]. One can note that the cutoff in the case of longer pulses (25th harmonic) was shorter with regard to the few-cycle pulses at similar intensities of these pulses in the plasma plume.

The problem with using a fullerene powder-containing target is the shot-to-shot instability and rapid decrease of the harmonic yield, due to the abrupt change in the target morphology following ablation. Such instabilities limit applications of this radiation source and greatly hamper efforts to measure the pulse duration of the harmonic emission, which typically requires a large number of laser shots. A solution to this problem might be the use of long homogeneous tapes containing fullerenes, which continuously move from shot to shot to provide a fresh surface for each next laser pulse. The use of rotating targets containing C60 is another method for improvement of the harmonic stability ([16], see also previous section) from this medium, which was implemented in the discussed work. The 15 mm diameter aluminum rods were used as the substrates onto which the fullerene powder was glued. These rotating targets considerably improved the stability of harmonics from C60-containing plasmas compared with fixed targets fabricated by gluing fullerene powder to a glass substrate. Once the target rotation is stopped, the harmonic efficiency from the fullerene plasma decreased to the noise level within 1 - 2 seconds.

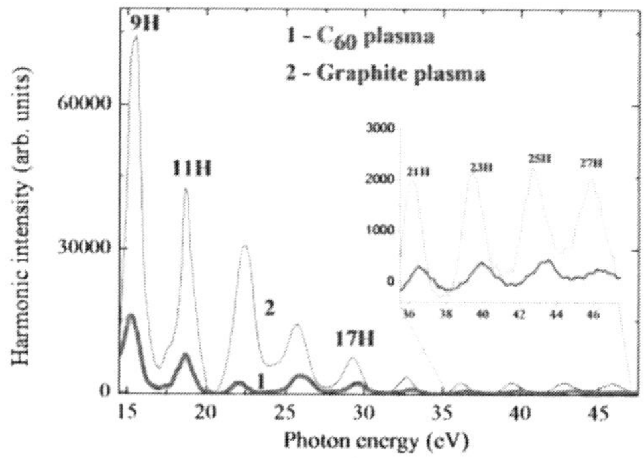

(a)

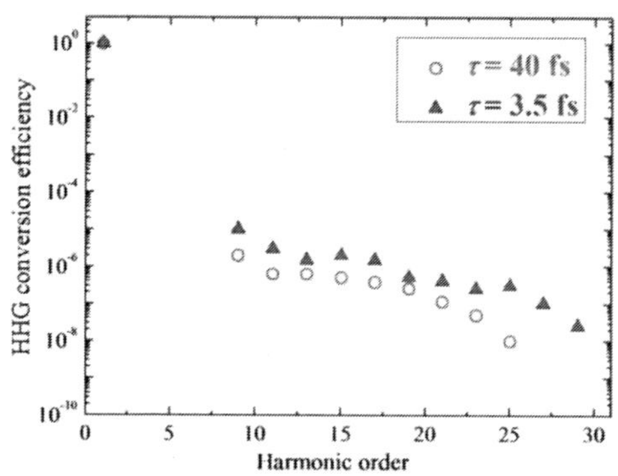

(b)

Figure 12. (a) Harmonic spectra from fullerene (thick curve) and graphite (thin curve) plasmas using the 8 ps heating pulses and 3.5 fs driving pulses at identical experimental conditions; (b) Comparison of HHG conversion efficiencies from fullerene plasma using the 3.5 fs (triangles) and 40 fs (circles) driving pulses. The laser wavelength is λ= 780 nm [52].

This target fabrication method could be very useful in the case of powder-like targets (fullerenes, metal nanoparticles, organic powder-like samples, etc.). Previously, even at 10 Hz pulse repetition rate, the ablation of powdered targets that were not moved led to their rapid degradation and abrupt decrease of harmonic yield [43]. With this new technique, one can glue the powdered material onto the rotating rod to improve the stability of the harmonic emission. At the same time, one can note that the stability of harmonics from such powder-like targets is still worse than from rotating bulk metal rods due to the faster degradation of the target surface in the former case.

Due to small concentration of fullerenes in the plasma plume, the amount of ablated material was insignificant even at 1 kHz pulse repetition rate. So the main difference between the rotating and non-rotating targets was related with the thermal conditions of heating spot. In the case of fixed target, the melting bath appeared after 1000 shots (e.g. 1 sec of ablation), which considerably worsened the process of plasma formation. Once the target started to move, the previously heated area cooled down and again could be used for efficient ablation and harmonic generation. This phenomenon was observed from many targets. It is possible to improve the stability of harmonics (and correspondingly stability of plasma formation conditions) not just by using the fresh (non-ablated) surface but by changing the conditions of overheating of the same spot of the target.

Conversion efficiency studies in these two plasmas showed advantages of HHG in the case of graphite plasmas compared with fullerene plasmas [37]. This can be explained by the higher particle density in the graphite plasma. The concentration of fullerenes is below 1017 cm−3 [46], while density estimates for carbon plasmas based on a three-dimensional molecular dynamical simulation of laser ablation of graphite using the molecular dynamics code ITAP IMD [53] showed that for heating by 8 ps laser pulses the graphite plasma density can reach 2.6 × 1017 cm−3 at the ablation intensity of 2×1010 W·cm−2. Another reason for the observed superior features of graphite plasma harmonics could be the production of clusters during laser ablation, though their involvement in HHG requires additional studies, including time-of-flight measurements.

In order to analyze the harmonic yield from the fullerene plasma using the mid infrared (MIR) laser source, an optical parametric oscillator (OPA) pumped by the 40 fs Ti:sapphire laser was used. A beam splitter inserted before the laser compressor of this Ti:sapphire laser was used to pick off 10% of the beam (780 nm, 1 mJ, 160 ps, 1 kHz pulses) to generate a fullerene-containing plasma plume, with the remaining 90% being compressed to 40 fs (7 mJ) to pump a computer-controlled OPA. The OPA was optimized for high conversion efficiency, beam quality and short duration of the converted pulses. The OPA provided signal pulses with duration of 35 fs in the 1200 - 1600 nm spectral range with a maximum energy of 1.7 mJ at ~1300 nm. The idler pulses covered the 1600 - 2200 nm range with a maximum energy of 1 mJ at ~ 2000 nm. The delay between the heating ablation pulse and MIR pulse from the OPA was set to 35 ns, as this delay was found to be optimal for the efficient generation of extended harmonics from fullerene plasma.

Figure 13 shows the comparison of fullerene harmonic spectra generated in the case of 1300 and 780 nm multi-cycle pulses. Harmonics up to the 41st order (Figure 13, bottom panel) were observed in the case of 1300 nm pulses at the conditions of optimal plasma formation using the heating 160 ps pulses. The application of less intense, longer wavelength (1400 nm) pulses available by tuning the OPA did not result in an extension of harmonic cut-off compared with the case of 1300 nm pulses. This observation suggests that the harmonic generation occurred under saturated conditions, with the expectation of stronger harmonics once the microand macro-processes governing frequency conversion are optimized. Over-excitation of target by 160 ps pulses (Figure 13, middle panel) led to appearance of plasma emission in the 25 - 45 eV range of photon energies (λ = 27 - 50 nm). At these conditions, no harmonics were observed during propagation of femtosecond pulses through such overexcited plasma.

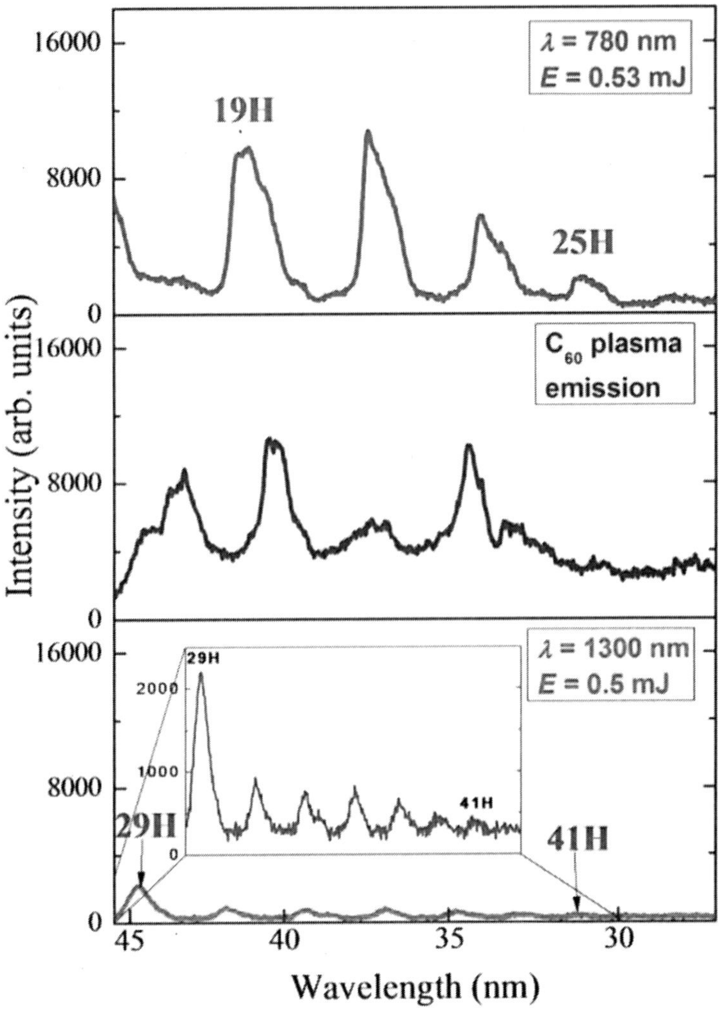

Figure 13. Comparative harmonic spectra from fullerenecontained plume using the 780 nm (upper panel) and 1300 nm (bottom panel) multi-cycle pulses (intensity of 160 ps heating pulses Ips = 1 × 1010 W·cm−2). The intensity axes are on the same scale allowing a direct comparison between the three cases. The middle panel shows C60 plasma emission spectrum at over-excitation of target by 160 ps heating pulses (Ips = 4 × 1010 W·cm−2), without further excitation by femtosecond driving pulses [52].

The harmonic spectrum up to the 25th order in the case of 780 nm, 40 fs pulses is presented in the Figure 13 (upper panel). One can clearly see the extension of harmonic cut-off (from the point of view of highest harmonic order) in the case of longer-wavelength driving pulses by comparing the HHG spectra using the 780 and 1300 nm pulses, while the extension of cut-off energy was insignificant.

Approximately equal energies of driving pulses were maintained in these cases (0.53 mJ for 780 nm pulses and 0.5 mJ for 1300 nm pulses). The intensity of 780 nm radiation in the plasma area was calculated to be Ifs = 4 × 1014 W·cm−2. The pulse durations of these sources (780 and 1300 nm) were approximately same (40 and 35 fs respectively). Probably due to phase modulation and propagation through multiple optical elements in optical parametric oscillator-amplifier, the diameter of 1300 nm beam in the plasma plume was larger than in the case of 780 nm beam. The corresponding lower intensity of 1300 nm radiation could be responsible for less expected extension of harmonic cut-off energy. From the cut-off formula, one could expect the generation of harmonics up to the 47th order (in the case of 780 nm radiation), well above the observed cut-off (25th harmonic), which probably points out the difference in expected and actual intensity in the plasma area at the optimal conditions of HHG. The reason in this discrepancy could be related with the self-defocusing of driving pulses in the medium containing free electrons.

In atoms the maximum emitted photon energy (giving the cut-off position) is described by the well known relation EM = I + 3.17Up with I the ionization energy and Up the ponderomotive energy. By invoking the three step model [44] this formula can be derived under the assumptions that the laser field amplitude is constant. Rapidly varying laser pulses can considerably shift the cutoff [54,55]. The molecules support more returning trajectories than atoms and allow the existence of several plateau [56]. It is not therefore easy to disentangle the effects of pulse shape and presence of a molecule and to state a general law for the position of the cutoff; the EM = I + 3.17Up law can therefore be used only as a touchstone.

It was observed that the plasma harmonic yields from the 780 and MIR probe pulses are consistent with the predicted single-atom harmonic intensity wavelength scaling Ih µ λ−5, which arises due to electron wavepacket spreading before recollision [57,58]. The harmonic efficiency of the XUV radiation in the range of 30 - 50 nm driven by MIR pulses was 7 to 15 times less compared with the case of 780 nm probe pulses, which is comparable with the expected ratio between harmonic intensities from these sources ((1300/780)5 » 12.7) followed from above rule, assuming approximately equal energies of the 780 and 1300 nm pulses (0.53 and 0.5 mJ respectively).

The fact that HHG in the fullerene system exhibits a wavelength scaling that is consistent with single-atom predictions allowed to employ a theoretical model [59] in which the interaction of short pulses with C60 was treated in the single active electron approximation with one electron constrained over a structureless spherical surface of radius R = 3.55 × 10−8 cm = 6.71a0. This simplified model can be treated in an analytical fashion. It permits a physical understanding of the dynamics of the active electron, and allows an easy check of resonances between bare molecular state. Indeed a generalization of the model can be used to analyze the physical properties, the energy content and stability of hollow spherical clusters [60]. The calculated harmonic spectra from C60 are presented in Figure 14 in the case of 780 nm (photon energy Eph = 1.6 eV) and 1300 nm (photon energy Eph = 0.96 eV) pulses propagating through the fullerene medium. The calculations were carried out for 2-cycle pulses (t = 5.2 fs) and 12-cycle pulses (t = 31 fs) of 780 nm radiation and 8- cycle pulses (t = 34 fs) of 1300 nm radiation and intensity 6 × 1014 W·cm−2, which were close to the conditions of fullerene HHG experiments.

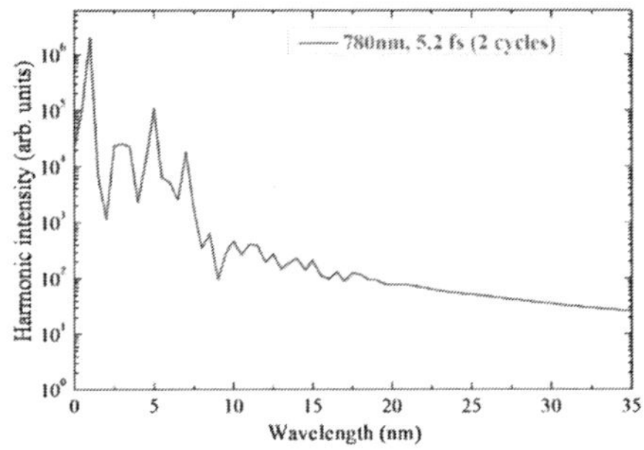

(a)

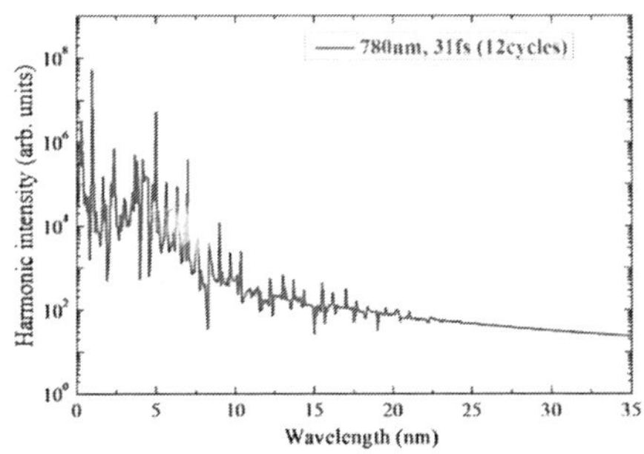

(b)

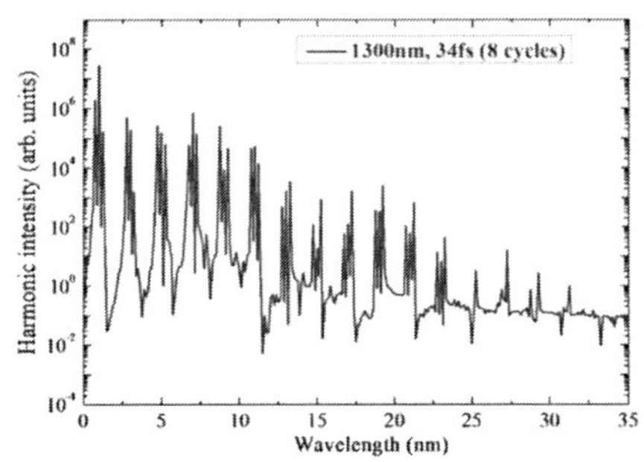

(c)

Figure 14. Calculated data of the harmonic spectra from fullerene plasma in the case of (a,b) 780 nm and (c) 1300 nm driving radiation. The pulse durations are (a) 2 optical cycles (5.2 fs), (b) 12 optical cycles (31 fs) and (c) 8 optical cycles (34 fs) [52].

The spectra of well-resolved harmonics were obtained but with broadened lines (in the case of short pulses) and hyper Raman lines (in the case of long pulses). Hyper Raman lines with frequency other than harmonics, are due to transitions between laser dressed molecular states [61,62]. The presence of these lines has been predicted since the very beginning of the theoretical treatment of HHG [63] and found in different contexts such as twolevel approximation, quantum dots calculations, hydrogen atom and so on [64-67] but never observed in actual experiments. Several explanations have been proposed to explain this failure; for example it has been argued that they add destructively in the forward direction or that they show a transient nature and are thus overwhelmed by the presented odd harmonics [68,69].

The calculations [52] showed well-defined harmonics (up to Hc = 31) in the case of 1300 nm multi-cycle pulses. The theoretical model exploited the spherical symmetry of the C60 by introducing a number of approximations, the most important of which is that the molecule cannot be ionized. This approximation deserves some comment. In spite of the ionization suppression of the C60 molecule, some ionization is bound to occur so that the theory becomes unreliable when ionization becomes significant. The harmonic cut-off in the case of 1300 nm radiation was extended compared with 780 nm radiation (Hc = 17), analogously as in the case of experiment.

Previous HHG studies in fullerenes were performed using the multi-cycle pulses (30 fs, 48 fs, and 110 fs [43]). The stability of C60 against fragmentation in multicycle laser fields leads to fast diffusion of the excitation energy. Even better conditions can occur in the case of few-cycle pulses used for fullerene HHG. In that case fullerenes can withstand the influence of the strong field of few-cycle pulses due to the increase of the ionization saturation intensity as the pulse duration is decreased. This can further increase the diffusion of the excitation energy within the fullerenes due to their very large number of internal degrees of freedom. The increase of energy diffusion is expected to lead to a decrease in the disintegration of fullerenes, which enhances the probability of harmonic emission from these molecules.

The graphite plasma, which was compared in these experiments with C60, can be easily aggregated during laser ablation, thus leading to nanoparticles formation in the laser plume, as it was shown in [36-38]. In that case the comparison of two clustered species (large 5 - 20 nm carbon clusters and 0.7 nm C60) can lead to their different nonlinear optical response once the interacting laser pulse becomes compressed from the multi-cycle to fewcycle duration. The discussed results show that for the shortest laser pulses, the HHG cut-off for fullerene is extended, while the harmonic intensity is reduced compared with "carbon harmonics". The intensity of the femtosecond pulse is an important parameter in for optimizing the HHG from C60. Increasing the intensity of the driving pulse did not lead to an extension of the harmonic cutoff from the fullerene plume, which is a signature of HHG saturation in the medium. Moreover, at relatively high intensities, a decrease in the harmonic output was observed, which can be attributed to the phase mismatch caused by propagation effects.

The stability of C60 molecules against ionization and fragmentation is of particular importance, especially for their application as a medium for HHG using laser pulses of different duration. The fullerene molecules ablated off the surface should be intact when the driving pulse arrives. Hence, the heating pulse intensity also becomes a sensitive parameter. At lower intensities of heating pulse, the concentration of the clusters in the plume would be low, while at higher intensities one can expect fragmentation. This phenomenon is observed when the heating pulse intensity on the surface of fullerene-rich targets is increased above the critical value (Figure 13, middle panel). The abrupt reduction in harmonic intensity in that case can be attributed to phenomena such as fragmentation of fullerenes, an increase in free electron density, and self-defocusing of the driving pulse, all of which are expected to reduce the efficiency of HHG.

Summarizing this section, the experimental and theoretical studies of high-order harmonic generation in the plasmas containing fullerenes (up to the 29th and 41st orders for the 780 and 1300 nm radiation) under different plasma conditions and laser parameters in the case of fewand multi-cycle pulses show the attractiveness of applications of this medium for frequency conversion of laser radiation. The comparison of harmonics from C60-rich plasmas and plasmas produced on graphite surface showed stronger harmonic yield in the latter case, which was attributed to higher density of the plasma plume and the formation of clusters during the ablation of graphite targets. The experiments with fullerene powder glued onto aluminum rotating rods demonstrated a dramatic improvement in harmonic generation stability compared with static fullerene-containing targets. The comparative studies using 3.5 and 40 fs pulses showed that, for few-cycle pulses, the harmonic cutoff is extended compared with multi-cycle pulses, which can be attributed to reduced fragmentation of C60 for the shorter pulse. The comparison of fullerene harmonic spectra generated in the case of 1300 and 780 nm multi-cycle pulses showed the extension of generating harmonic orders in the former case. The plasma harmonic yield approximately follows the Ih µ λ–5 wavelength scaling rule predicted for single atoms. The maximal conversion efficiency obtained in these fullerene HHG studies was estimated to be 5 × 10–6. Theoretical calculations of fullerene harmonic spectra were carried out in the single active electron approximation, which showed harmonics up to Hc = 31 in the case of 1300 nm multi-cycle pulses and up to Hc = 17 in the case of 780 nm multi-cycle pulses.

6. PERSPECTIVES OF PLASMA HHG

The quest for new plasma media that would favor the enhancement of an individual harmonic allows further enhancement of harmonic conversion efficiency. The production of a single high-intensity harmonic (rather than a group of harmonics of equal intensity in the plateau region) would open up the way to the practical application of these coherent short-wavelength radiation sources. Because resonantly enhanced harmonics have already been observed in several plasma media, there are strong grounds to believe that similar conditions will be discovered for other plasma formations. The generated harmonic wavelength may then be tuned to the transitions with high oscillator strength by wavelength tuning of the driving laser system, as well as by varying the chirp of the laser radiation. Application of ablated nanoparticles and clusters for HHG can also enhance the yield of harmonics in the XUV range. Further improvements in HHG conversion efficiency and harmonic extension require a systematic study of the influence of various plasma and laser parameters on ablation harmonics. Many new peculiarities of plasma harmonics emerged during last few years allow expecting further extension of our knowledge of materials properties using this powerful tool of nonlinear spectroscopy.

Whilst the first stage of these successful studies was entirely focused on the improvements of harmonic yield from plasma, at current stage of knowledge of the high-order nonlinear optical processes in ablation plume one can consider this method as a new technique for materials science. Thus the search of the dual role of plasma HHG as a method for efficient coherent XUV light generation and of materials probing is a milestone for future developments and the next stage of studies would be aimed at the enhancement of HHG efficiency from laser ablation produced on the surfaces of solid-state materials and applications of plasma harmonics for analysis of spectral and structural properties of materials.

Future studies are aimed on an analysis of the HHG spectroscopy conducted using a low-excited laser-produced plasma, which has already demonstrated several new approaches to the problem of increasing the HHG efficiency in comparison with the efficiency of conventional HHG in gases. At the same time, it is obvious that gas HHG exhibits much higher harmonic orders (in comparison with the plasma HHG obtained to date). Without making excessively optimistic predictions that the orders of in-plasma-generated harmonics will exceed the in-gas harmonic orders in the nearest future, the researchers can focus primarily on the realization of new techniques for increasing the wavelength-converted radiation intensity with the use of plasma plumes, which may hardly be realized in conventional gas HHG. At the same time, these studies will allow to explore various optical, nonlinear optical, and structural properties of materials using the laser ablation induced HHG spectroscopy.

It is worth noting that, currently, nonlinear spectroscopy involving gas HHG trying to deal with such problems as (a) extension of the HHG spectroscopy based attosecond structural imaging technique to image nuclear re-arrangements induced by localized hole excitations, (b) application of strong-field ionization to create localized hole excitations and study their attosecond dynamics in polyatomic molecules, (c) search of selective imaging of hole dynamics induced by the removal of e.g. inner-valence electrons using the XUV initiated HHG technique, and (d) development of multi-dimensional HHG spectroscopy capable of following energy flow between different molecular modes over multiple femtosecond timescale. The proposed laser ablation induced HHG spectroscopy allows adding some important impulse to those studies, by using the peculiarities of plasma HHG.

One can expect an improvement of plasma HHG due to the double excitation of laser-produced plasmas, optimization of the longitudinal harmonic generation scheme in the laser plume, optimization of nanostructured plasmas, use of multicomponent plasma plumes, formation of quasi-phase-matching conditions for the waves in complex extended plasmas, provision of a regime of waveguide pump propagation through the plasma medium, feasibility analysis of atto second pulse generation in laser-produced plasma, use of mid-IR pulses for harmonic extension, stabilization of plasmas and harmonics characteristics using rotating and moving targets, search of the attosecond pulse generation by achieving the continuum in the harmonic emission near the cut-off, which allows for the first time the generation of ultrashort laser pulses in the plasma plumes, etc. All these concepts are new and did not apply previously for the improvement of plasma harmonic generation efficiency. The joint implementation of new and old (application of the clusters with controllable and variable sizes, resonance-induced harmonic enhancement, two-color pump-induced enhancement of even harmonics, search of the influence of multi-electron dynamics of complex clusters, such as nanotubes and graphenes, on the plasmon resonanceinduced growth of few harmonics in the XUV range, etc.) methods and approaches in the case of plasma media allows making the advance to the state-of-the-art within the plasma HHG field, which will lead to the establishment of a new method of material science laser ablation induced high-order harmonic generation spectroscopy. It follows from the above that investigations in this area of nonlinear optics are making rapid strides and may bring new success in the nearest future.

7. ACKNOWLEDGEMENTS

Author thanks J. P. Marangos, A. V. Andreev, H. Kuroda, S. Y. Stremoukhov, T. Witting, C. Hutchison, F. Frank, M. Tudorovskaya, M. Lein, A. Zaïr, J. W. G. Tisch, T. Ozaki, and E. Fiordilino for fruitful discussions and collaboration.

REFERENCES

1. L N. A. Papadogiannis, L. A. A. Nikolopoulos and D. Charalambidis, "Two-Photon Ionization of He through a Superposition of Higher Harmonics," Physical Review Letters, Vol. 90, No. 13, 2003, Article ID: 133902.
2. J. Itatani, J. Levesque, D. Zeidler, H. Niikura, H. Pépin, J.-C. Kieffer, P. B. Corkum and D. M. Villeneuve, "Tomographic Imaging of Molecular Orbitals," Nature, Vol. 432, No. 7019, 2004, pp. 867-872.
3. R. I. Tobey, M. E. Siemens, O. Cohen, M. M. Murnane, H. C. Kapteyn and K. A. Nelson, "Ultrafast Extreme Ultraviolet

Holography: Dynamic Monitoring of Surface Deformation," Optics Letters, Vol. 32, No. 3, 2007, pp. 286-288.

4. P. M. Paul, E. S. Toma, P. Breger, G. Mullot, F. Auge, P. Balcou, H. G. Muller and P. Agostini, "Observation of a Train of Attosecond Pulses from High Harmonic Generation," Science, Vol. 292, No. 5522, 2001, pp. 1689- 1692. doi:10.1126/science.1059413

5. M. Drescher, M. Hentschel, R. Kienberger, G. Tempea, C. Spielmann, G. A. Reider, P. B. Corkum and F Krausz, "X-Ray Pulses Approaching the Attosecond Frontier," Science, Vol. 291, No. 5510, 2001, pp. 1923-1927.

6. P. B. Corkum and F. Krausz, "Attosecond Science," Nature Physics, Vol. 3, No. 6, 2007, pp. 381-387.

7. P. V. Redkin and R. A. Ganeev, "Simulation of Resonant High-Order Harmonic Generation in a Three-Dimensional Fullerenelike System by Means of a Multiconfigurational Time-Dependent Hartree-Fock Approach," Physical Review A, Vol. 81, No. 6, 2010, Article ID: 063825.

8. R. A. Ganeev, H. Singhal, P. A. Naik, U. Chakravarty, V. Arora, J. A. Chakera, R. A. Khan, M. Raghuramaiah, S. R. Kumbhare, R. P. Kushwaha and P. D. Gupta, "Optimization of the High-Order Harmonics Generated from Silver Plasma," Applied Physics B, Vol. 87, No. 2, 2007, pp. 243- 247.

9. R. A. Ganeev, L. B. Elouga Bom, J.-C. Kieffer and T. Ozaki, "Demonstration of the 101st Harmonic Generated from a Laser-Produced Manganese Plasma," Physical Review A, Vol. 76, No. 2, 2007, Article ID: 023831.

10. R. A. Ganeev, M. Baba, M. Suzuki and H. Kuroda, "High-Order Harmonic Generation from Silver Plasma," Physics Letters A, Vol. 339, No. 1-2, 2005, pp. 103-106.

11. R. A. Ganeev, "Generation of High-Order Harmonics of High-Power Lasers in Plasmas Produced Under Irradiation of Solid Target Surfaces by a Prepulse," Physics Uspekhi, Vol. 52, No. 1, 2009, pp. 55-79.

12. M. Suzuki, M. Baba, R. A. Ganeev, H. Kuroda and T. Ozaki, "Intense Exact Resonance Enhancement of Single-High-Harmonic from an Antimony Ion by Using Ti: Sapphire Laser at 37 nm," Optics Express, Vol. 15, No. 3, 2007, pp.1161-1166.
13. A. V. Andreev, R. A. Ganeev, H. Kuroda, S. Y. Stremoukhov and O. A. Shoutova, "High-Order Harmonic Cut-Off Frequency in Atomic Silver Irradiated by Femtosecond Laser Pulses: Theory and Experiment," European Physical Journal D, Vol. 67, No. 1, 2013, p. 22.
14. V. Andreev, S. Y. Stremoukhov and O. A. Shoutova, "Nonlinear Optical Response of an Atom in the Field of Femtosecond Laser Pulses with Near-Atomic Intensity," JETP Letters, Vol. 93, No. 8, 2011, pp. 476-486.
15. R. A. Ganeev, "High-Order Harmonic Generation in a Laser Plasma: a Review of Recent Achievements," Journal of Physics B, Vol. 40, No. 22, 2007, pp. R213-R254.
16. Hutchison, R. A. Ganeev, T. Witting, F. Frank, W. A. Okell, J. W. G. Tisch and J. P. Marangos, "Stable Generation of High-Order Harmonics of Femtosecond Laser Radiation from Laser Produced Plasma Plumes at 1 kHz Pulse Repetition Rate," Optics Letters, Vol. 37, No. 11, 2012, pp. 2064-2066.
17. A. V. Andreev, "Interaction of an Atom with Superstrong Laser Fields," Journal of Experimental and Theoretical Physics, Vol. 89, No. 3, 1999, pp. 421-427.
18. D. Shiner, B. E. Schmidt, C. Trallero-Herrero, H. J. Wörner, S. Patchkovskii, P. B. Corkum, J.-C. Kieffer, F. Légaré and D. M. Villeneuve, "Probing Collective MultiElectron Dynamics in Xenon with High-Harmonic Spectroscopy," Nature Physics, Vol. 7, No. 6, 2011, pp. 464- 467.
19. R. A. Ganeev, M. Suzuki, T. Ozaki, M. Baba and H. Kuroda, "Strong Resonance Enhancement of a Single Harmonic Generated in the Extreme Ultraviolet Range," Optics Letters, Vol. 31, No. 11, 2006, pp. 1699-1701.
20. R. A. Ganeev, P. A. Naik, H. Singhal, J. A. Chakera and P. D. Gupta, "Strong Enhancement and Extinction of Single Harmonic

Intensity in the Midand End-Plateau Regions of the High Harmonics Generated in Weakly Excited Laser Plasmas," Optics Letters, Vol. 32, No. 1, 2007, pp. 65-67.

21. R. A. Ganeev, M. Suzuki, M. Baba and H. Kuroda, "Extended High-Order Harmonics from Laser-Produced Cd and Cr Plasmas," Applied Physics Letters, Vol. 94, No. 5, 2009, Article ID: 051101.
22. M. Hentschel, R. Kienberger, C. Spielmann, G. A. Reider, N. Milosevic, T. Brabec, P. Corkum, U. Heinzmann, M. Drescher and F. Krausz, "Attosecond Metrology," Nature, Vol. 414, No. 6863, 2001, pp. 509-514.
23. Altucci, J. W. G. Tisch and R. Velotta, "Single Attosecond Light Pulses from Multi-Cycle Laser Sources," Journal of Modern Optics, Vol. 58, No. 18, 2011, pp. 1585-1610.
24. T. Witting, F. Frank, W. A. Okell, C. A. Arrell, J. P. Marangos and J. W. G. Tisch, "Sub-4-fs Laser Pulse Characterization by Spatially Resolved Spectral Shearing Interferometry and Attosecond Streaking," Journal of Physics B, Vol. 45, No. 7, 2012, Article ID: 074014.
25. R. A. Ganeev, T. Witting, C. Hutchison, F. Frank, M. Tudorovskaya, M. Lein, W. A. Okell, A. Zaïr, J. P. Marangos and J. W. G. Tisch, "Isolated sub-fs XUV Pulse Generation in Mn Plasma Ablation," Optics Express, Vol. 20, No. 23, 2012, pp. 25239-25248.
26. Kilbane, E. T. Kennedy, J. P. Mosnier, J. T. Costello and P. van Kampen, "EUV Photoabsorption Spectra of Cd II and Cd III," Journal of Physics B, Vol. 38, No. 1, 2005, pp. 83-88.
27. V. K. Dolmatov, "Characteristic Features of the 3p Absorption Spectra of Free Iron-Group Elements due to the Duplicity of the 'Inner-Valence' 3d Electrons. Application to Mn2+," Journal of Physics B, Vol. 29, No. 29, 1996, pp. L687-L692. doi:10.1088/0953-4075/29/19/001
28. H. Kjeldsen, F. Folkmann, B. Kristensen, J. B. West and J. E. Hansen, "Absolute Cross Section for Photoionization of Mn+ in

the 3p Region," Journal of Physics B, Vol. 37, No. 6, 2004, pp. 1321-1330.

29. J. B. West, J. E. Hansen, B. Kristensen, F. Folkmann and H. Kjeldsen, "Revised Interpretation of the Photoionization of Cr+ in the 3p Excitation Region," Journal of Physics B, Vol. 36, No. 19, 2003, pp. L327-334.
30. V. Strelkov, "Role of Autoionizing State in Resonant High-Order Harmonic Generation and Attosecond Pulse Production," Physical Review Letters, Vol. 104, No. 12, 2010, Article ID: 123901.
31. M. V. Frolov, N. L. Manakov and A. F. Starace, "Potential Barrier Effects in High-Order Harmonic Generation by Transition-Metal Ions," Physical Review A, Vol. 82, No. 2, 2010, Article ID: 023424.
32. M. Tudorovskaya and M. Lein, "High-Order Harmonic Generation in the Presence of a Resonance," Physical Review A, Vol. 84, No. 1, 2011, Article ID: 013430.
33. R. A. Ganeev, C. Hutchison, T. Siegel, M. E. López-Arias, A. Zaïr and J. P. Marangos, "High-Order Harmonic Generation from Metal Plasmas Using 1 kHz Laser Pulses," Journal of Modern Optics, Vol. 58, No. 10, 2011, pp. 819-822.
34. R. A. Ganeev, C. Hutchison, T. Siegel, A. Zaïr and J. P. Marangos, "Quantum Path Signatures in Harmonic Spectra from Metal Plasma," Physical Review A, Vol. 83, No. 6, 2011, Article ID: 063837.
35. R. A. Ganeev, C. Hutchison, A. Zaïr, T. Witting, F. Frank, W. A. Okell, J. W. G. Tisch and J. P. Marangos, "Enhancement of High Harmonics from Plasmas Using TwoColor Pump and Chirp Variation of 1 kHz Ti:sapphire Laser Pulses," Optics Express, Vol. 20, No. 1, 2012, pp. 90-100.
36. L. B. Elouga Bom, S. Haessler, O. Gobert, M. Perdrix, F. Lepetit, J.-F. Hergott, B. Carré, T. Ozaki and P. Salières, "Attosecond Emission from Chromium Plasma," Optics Express, Vol. 19, No. 4, 2011, pp. 3677-3685.

37. Y. Pertot, L. B. Elouga Bom, V. R. Bhardwaj and T. Ozaki, "Pencil Lead Plasma for Generating Multimicrojoule High-Order Harmonics with a Broad Spectrum," Applied Physics Letters, Vol. 98, No. 10, 2011, p. 101104.
38. R. A. Ganeev, T. Witting, C. Hutchison, F. Frank, P. V. Redkin, W. A. Okell, D. Y. Lei, T. Roschuk, S. A. Maier, J. P. Marangos and J. W. G. Tisch, "Enhanced High-Order Harmonic Generation in a Carbon Ablation Plume," Physical Review A, Vol. 85, No. 1, 2012, Article ID: 015807.
39. W. Theobald, C. Wulker, F. R. Schafer and B. N. Chichkov, "High-Order Harmonic Generation in Carbon Vapor and Low Charged Plasma," Optics Communications, Vol. 120, No. 3-4, 1995, pp. 177-183.
40. M. López-Arias, M. Oujja, M. Sanz, R. A. Ganeev, G. S. Boltaev, N. K. Satlikov, R. I. Tugushev, T. Usmanov and M. Castillejo, "Low-Order Harmonic Generation in Metal Ablation Plasmas in Nanosecond and Picosecond Laser Regimes," Journal of Applied Physics, Vol. 111, No. 4, 2012, Article ID: 043111.
41. T. D. Donnelly, T. Ditmire, K. Neuman, M. D. Perry and R. W. Falcone, "High-Order Harmonic Generation in Atom Clusters," Physical Review Letters, Vol. 76, No. 14, 1996, pp. 2472-2475.
42. Vozzi, M. Nisoli, J.-P. Caumes, G. Sansone, S. Stagira, S. De Silvestri, M. Vecchiocattivi, D. Bassi, M. Pascolini, L. Poletto, P. Villoresi and G. Tondello, "Cluster Effects in High-Order Harmonics Generated by Ultrashort Light Pulses," Applied Physics Letters, Vol. 86, No. 11, 2005, Article ID: 111121.
43. R. A. Ganeev, "Harmonic Generation in Laser-Produced Plasmas Containing Atoms, Ions and Clusters: A Review," Journal of Modern Optics, Vol. 59, No. 5, 2012, pp. 409-439.
44. P. B. Corkum, "Plasma Perspective on Strong Field Multiphoton Ionization," Physical Review Letters, Vol. 71, No. 13, 1993, pp. 1994-1997.
45. S. V. Popruzhenko, D. F. Zaretsky and D. Bauer, "Energy Absorption and Emission of Harmonics by Clusters Subject to

Intense Short Laser Pulses," Laser Physics Letters, Vol. 5, No. 9, 2008, pp. 631-646.

46. R. A. Ganeev, L. B. Elouga Bom, J. Abdul-Hadi, M. C. H. Wong, J. P. Brichta, V. R. Bhardwaj and T. Ozaki, "HigherOrder Harmonic Generation from Fullerene by Means of the Plasma Harmonic Method," Physical Review Letters, Vol. 102, No. 1, 2009, Article ID: 013903.
47. M. F. Ciappina, A. Becker and A. Jaron-Becker, "Multislit Interference Patterns in High-Order Harmonic Generation in C60," Physical Review A, Vol. 76, No. 6, 2007, Article ID: 063406.
48. M. Ruggenthaler, S. V. Popruzhenko and D. Bauer, "Recollision-Induced Plasmon Excitation in Strong Laser Fields," Physical Review A, Vol. 78, No. 3, 2008, Article ID: 033413.
49. G. P. Zhang, "Optical High Harmonic Generation in C60," Physical Review Letters, Vol. 95, No. 4, 2005, Article ID: 047401. doi:10.1103/PhysRevLett.95.047401
50. H.-D. Meyer, U. Manthe and L. S. Cederbaum, "The Multi-Configurational Time-Dependent Hartree Approach," Chemical Physics Letters, Vol. 165, No. 1, 1990, pp. 73-78.
51. V. R. Bhardwaj, P. B. Corkum and D. M. Rayner, "Recollision during the High Laser Intensity Ionization of C60," Physical Review Letters, Vol. 93, No. 4, 2004, Article ID: 043001.
52. R. A. Ganeev, C. Hutchison, T. Witting, F. Frank, S. Weber, W. A. Okell, E. Fiordilino, D. Cricchio, F. Persico, A. Zaïr, J. W. G. Tisch and J. P. Marangos, "HighOrder Harmonic Generation in Fullerenes Using Fewand Multi-Cycle Pulses of Different Wavelengths," Journal of the Optical Society of America B, Vol. 30, No. 1, 2013, pp. 7-12.
53. J. Roth, F. Géahler and H.-R. Trebin, "A Molecular Dynamics, Run with 5 180 116 000 Particles," Journal of Modern Physics C, Vol. 11, No. 2, 2000, pp. 317-322.
54. Fiordilino and V. Miceli, "Laser Pulse Shape Effects in Harmonic Generation from a Two-Level Atom," Journal of Modern Optics, Vol. 41, No. 7, 1994, pp. 1415-1426.

55. J. Zhou, J. Peatross, M. M. Murnane, H. C. Kapteyn and I. P. Christov, "Enhanced High-Harmonic Generation Using 25 fs Laser Pulses," Physical Review Letters, Vol. 76, No. 5, 1996, pp. 752-755.
56. M. Lein, "Mechanisms of Ultrahigh-Order Harmonic Generation," Physical Review A, Vol. 72, No. 5, 2005, Article ID: 053816.
57. J. Tate, T. Auguste, H. G. Muller, P. Salières, P. Agostini and L. F. DiMauro, "Scaling of Wave-Packet Dynamics in an Intense Midinfrared Field," Physical Review Letters, Vol. 98, No. 1, 2007, Article ID: 013901.
58. K. Schiessl, K. L. Ishikawa, E. Persson and J. Burgdörfer, "Quantum Path Interference in the Wavelength Dependence of High-Harmonic Generation," Physical Review Letters, Vol. 99, No 25, 2007, Article ID: 253903.
59. Cricchio, P. P. Corso, E. Fiordilino, G. Orlando and F. Persico, "A Paradigm of Fullerene," Journal of Physics B, Vol. 42, No. 8, 2009, Article ID: 085404.
60. Cricchio, E. Fiordilino and F. Persico, "Electrons on a Spherical Surface: Physical Properties and Hollow Spherical Clusters," Physical Review A, Vol. 86, No. 1, 2012, Article ID: 013201.
61. N. Moiseyev and M. Lein, "Non-Hermitian Quantum Mechanics for High-Order Harmonic Generation Spectra," Journal of Phycial Chemistry A, Vol. 107, No. 37, 2003, pp. 7181-7188. doi:10.1021/jp034390y
62. Z.-Y. Zhou and J.-M. Yuan, "Strengthened Hyper-Raman Lines with the Coherent Superposition State," Chinese Physics Letters, Vol. 24, No. 3, 2007, pp. 683-686.
63. T. Millack and A. Maquet, "Hyper-Raman Lines Produced During High Harmonic Generation," Journal of Modern Optics, Vol. 40, No. 11, 1993, pp. 2161-2171.
64. I. Gauthey, C. H. Keitel, P. L. Knight and A. Maquet, "Role of Initial Coherence in the Generation of Harmonics and Sidebands from a Strongly Driven Two-Level Atom," Physical Review A, Vol. 52, No. 1, 1995, pp. 525- 540.

65. W. Chu, Y. Xie, S. Duan, N. Yang, W. Zhang, J.-L. Zhu, and X.-G. Zhao, "Model of Generating either Odd or Even Optical Harmonics by Varying the Coupling Parameters Between Source Quantum Dots," Physical Review B, Vol. 82, No. 12, 2010, Article ID: 125301.

66. Z.-Y. Zhou and J.-M. Yuan, "Fine Structures of the Harmonic and Hyper-Raman Spectrum of the Hydrogen Atom in an Intense High-Frequency Laser Pulse," Physical Review A, Vol. 77, No. 6, 2008, Article ID: 063411.

67. V. Kapoor and D. Bauer, "Floquet Analysis of Real-Time Wave Functions without Solving the Floquet Equation," Physical Review A, Vol. 85, No. 2, 2012, Article ID: 023407.

68. A Di Piazza and E. Fiordilino, "Why Hyper-Raman Lines Are Absent in High-Order Harmonic Generation," Physical Review A, Vol. 64, No. 1, 2001, Article ID: 013802.

69. D. Bandrauk, S. Chelkowski and H. S. Nguyen, "Nonlinear Photon Processes in Molecules at High Intensities-Route to XUV—Attosecond Pulse Generation," Journal of Molecular Structure, Vol. 735-736, No. 2, 2005, pp. 203-209.

Chapter 2

Gain Coefficient Calculation for Short Wave Laser Emission from Sodium like Co

Wessameldin S. Abdelaziz[1], Mai E. Ahmed[2], Mohamed Atta Khedr[1], Tharwat M. El-Sherbini[3]

[1]National Institute of Laser Enhanced Sciences, Cairo University, Cairo, Egypt
[2]Environmental Affairs Agency, Cairo, Egypt
[3]Laboratory of Lasers and New Materials, Cairo University, Cairo, Egypt

ABSTRACT

Level structure, oscillator strengths, transition probabilities and radiative life times are evaluated for $1s^2\ 2s^2\ 2p^6 3l$, $4l$, $5l$ ($l = 0, 1, 2, 3, 4$) states in sodium like Co^{16+}. The calculations are carried out using COWAN code. The calculations made were compared with other results in literature where a good agreement is found. We also report on some unpublished energy values and oscillator strengths. Our results are used in the calculation of reduced population of 21 fine structure levels over a wide rang of electron density values (10^{18} to 10^{20}) at various electron plasma temperature. For those transitions with positive population inversion factor, the gain coefficients are evaluated and plotted against the electron density.

Keywords: Level Structure; Oscillator Strengths; Transition Probabilities and Radiative Life Times; Sodium like Co^{16+}

1. INTRODUCTION

An important objective in the development of X-ray lasers is to deliver a coherent, saturated output at wavelengths toward the water window [1].

Such saturated X-ray lasers are required for holography [2] and microscopy [3], of biological specimens and for deflectometry [4], interferometry [5], and radiography [6] of dense plasma relevant to inertial confinement fusion and laboratory astrophysics [5]. Sodium-like ions have prominent emission lines in the UV and XUV spectrum of the sun. Highly charged sodium-like ions are observed in several types of laboratory sources such as high-voltage vacuum spark tokamak and laser-produced plasmas [1].

Spectra of these ions have a simple structure (one electron outside a closed shell); the energy levels are practically free from effects of configuration mixing and therefore they are well suited for a theoretical interpretation of line intensities and for diagnostic purposes. In recent years, there have been extensive spectroscopic studies, both experimental and theoretical, of sodium isoelectronic sequence. [7] has used multi-configuration Hartree Fock (MCHF) method and Non-Orthogonal Spline configuration interaction (CI) method to evaluate energy levels and oscillator strengths for Na-like ions up to Fe XVI. [8] have used theoretical single-configuration Dirac-Fock method to evaluate oscillator strengths for E1 transitions in the sodium isoelectronic sequence (Na I-Ca X). Large-scale calculations were undertaken within the so-called Opacity Project (OP) [8] and, as a result of broad international collaboration, a complete set of oscillator strengths was produced for all optically allowed transitions between states with $1 \leq Z \leq 14$ as well as $Z = 16, 18, 20$, and 26 in all stages of ionization.

However, relativistic effects were also neglected in OP calculations, and LS coupling was assumed. Therefore, the OP data are available for multiplets only and not for individual fine-structure components. [9] assigned experimental wavelengths to 1086 lines which have oscillator strengths calculated in OP and added 1163 lines which have critically evaluated oscillator strengths. [10] and, from the US National Bureau of Standards (NBS) publications [11-13]. Experimental lifetime measurements for the 4p and 5p levels and calculation of transition probabilities in Na I have been measured using time-resolved laser induced fluorescence [14].

The purpose of this work is to present the energies of 21 fine structure levels, the oscillator strengths, the transition probabilities between them in sodium like Co^{16+}.

The atomic data thus obtained are used to calculate reduced population of sodium like Co^{16+} excited levels over a wide range of electron density at various electron temperatures. The gain coefficients are also calculated.

2. COMPUTATION OF ATOMIC STRUCTURES

2.1. Model of Central Force Field

In quantum mechanics, various physical processes can be summed by Schrödinger equation, i.e.

$$H\Psi_i = E_i\Psi_i. \tag{1}$$

In the non-relativistic case (the influence of relativistic effect will be discussed later), the Hamiltonian of an atomic system with N electrons is [1]:

$$\begin{aligned} H &= H_{kin} + H_{e\text{-}nuc} + H_{e\text{-}e} \\ &= \sum_i \frac{(h/2\pi)^2}{2m_e}\nabla_i^2 - \sum_i \frac{Ze^2}{r_i} + \sum_{i>j}\frac{e^2}{r_{ij}}. \end{aligned} \tag{2}$$

Here H_{kin}, $H_{e\text{-}nuc}$ and $H_{e\text{-}e}$ refer, respectively, to the kinetic energy of electrons, the Coulomb potential and the energy of electrostatic interaction of electrons, r_i is the distance between the i-th electron and nucleus, and $r_{i,j} = |r_i - r_j|$.

By substituting the Hamiltonian into Schrödinger equation and solving the equation in the case of multiple electrons and multiple energy levels, the wave function is obtained. Now, due to the appearance of the term of interaction of electrons, an exact solution cannot be obtained. On the other hand, the interaction term is comparable with the Coulomb potential term, so it can by no means be ignored. An approximate solution is to adopt the method of central force field. If it is assumed that every electron moves in the central force field of the nucleus and also in the mean force field produced by other electrons, then we have the following effective Hamiltonian [1]:

$$H^{eff} = \sum_{i}^{N} H_i^{eff} = -\sum_{i=1}^{N} \left[\frac{1}{2} \frac{p_i^2}{m_e} + \frac{Ze^2}{r_i} - V_i^{eff}(r_i) \right]. \qquad (3)$$

where Z is the atomic number

2.2. Method of Calculation

The key problem in the application of central field is to find an adequate potential function V^{eff}. For this, in recent decades many effective method of calculation have been developed. Among them the more important ones are the potential model, Hartree-Fock theory, the semi-empirical methods. In the following we present a brief introduction of semi-empirical methods.

Semi-empirical methods try to calculate atomic structures via solving the simplified form of the Hartree-Fock equation. The most typical is the Hartree-Fock-Slater method.

Afterwards, Cowan et al. revised this method and developed the RCN/RCG program used in our work. The merit of the program is its extreme effectiveness, and the shortcoming is its inability to estimate the precision.

2.3. Configuration Interaction

In the above-stated model of central force field, every electron can be described with a simple wave function. The overall wave function of atoms may be expressed with the following Slater determinant [1]:

$$\Phi = \frac{1}{\sqrt{N!}} \begin{pmatrix} \varphi_1(\chi_1) & \mathrm{K} & \varphi_1(\chi_N) \\ \mathrm{M} & \mathrm{O} & \mathrm{M} \\ \varphi_N(\chi_1) & \mathrm{L} & \varphi_N(\chi_N) \end{pmatrix}. \tag{4}$$

Φ: is a total function with respect to exchange of electrons.

ϕ(χ): is a single particle function or spin orbitals.

N: is a number of non-interacting electrons with no spin-orbit interaction.

In reality, such a description is not very precise. The best wave function should be a linear combination of wave functions with single configurations, and these wave functions possess the same total angular momentum and spin symmetry. This method is called the interaction of configurations. In the computation of atomic structures, consideration of the configuration interaction is the basis requirement for a program.

2.4. Relativistic Correction

In a non-relativistic system, the oscillator strengths and dipole transitions under LS-coupling can be calculated. In calculating forbidden transitions, jj-coupling must be used, and for this relativistic effects have to be taken into account. Generally speaking, the effects may be treated in two ways. One is inclusion of Breit-Pauli operator in the non-relativistic equation, and other is direct solution of the Dirac equation. For the former, a mass velocity term, the Darwin term caused by the electric moments of electrons and the spin-orbit term are added to the Hamiltonian of the model of central force field [15]. For relativistic correction, the program RCN/RCG restore to the Breit-Pauli correction.

2.5. Weighted Oscillator Strengths and Lifetimes

The oscillator strength f(γγ\) is a physical quantity related to line intensity I and transition probability W(γγ\), as given by Sobelman [16]:

$$W\left(\gamma\gamma^{\backslash}\right)=\frac{2w^{2}e^{2}}{mc^{3}}\left|f\left(\gamma\gamma^{\backslash}\right)\right| \tag{5}$$

with, I α gW(γγ\) α g|f(γγ\)| = gf.

Here m = electron mass, e = electron charge, = is initial quantum state, $W=\left(E(\gamma)-E\left(\gamma^{\backslash}\right)\right)/\mathrm{h}$, E(γ) initial state energy, g = (2J + 1) is the number of degenerate quantum state with angular momentum J (in the formula for initial state). Quantities with primes refer to the final state.

In the above equation, the weighted oscillator strength, gf, is given by Cowan [17]:

$$gf=\frac{8\pi^{2}mca_{0}^{2}\sigma}{3h}S, \tag{6}$$

where g is the statistical weight of lower level, f is the absorption oscillator strength, $\sigma=\left(E(\gamma)-E\left(\gamma^{\backslash}\right)\right)/hc$, h is planck's constant, c = light velocity, and a_0 is Bohr radius, and the electric dipole line strength is defined by:

$$s=\left|\left\langle\lambda J\left\|P^{1}\right\|\gamma^{\backslash}J^{\backslash}\right\rangle\right|^{2}. \tag{7}$$

This quantity is a measure of the total strength of the spectral line, including all possible transition between m, m' for different J_z Eigen states. The tensor operator P^1 (first order) in the reduced matrix element is the classical dipole moment for the atom in units of ea_0.

To obtain gf, we need to calculate S first, (or its square root):

$$s_{\gamma\gamma^{\backslash}}^{1/2} = \left\langle \lambda J \left\| P^1 \right\| \gamma^{\backslash} J^{\backslash} \right\rangle \tag{8}$$

In a multi-configuration calculation we have to expand the wave-function $|\gamma J\rangle$.

In terms of single configuration wave-functions, $|\beta J\rangle$ for both upper and lower levels:

$$|\gamma J\rangle = \sum_{\beta} y_{\beta J}^{\gamma} \, |\beta J\rangle. \tag{9}$$

therefore, we can have the multiconfigurational expression for the square root of line strength:

$$s_{\gamma\gamma^{\backslash}}^{1/2} = \sum_{\beta} \sum_{\beta^{\backslash}} y_{\beta J}^{\gamma} \left\langle \beta J \left\| P^1 \right\| \beta^{\backslash} J^{\backslash} \right\rangle \tag{10}$$

The probability per unit time of an atom in specific state γJ to make a spontaneous transition to any state with lower energy is

$$P(\gamma J) = \sum A(\gamma J, \gamma^{\backslash} J^{\backslash}) \tag{11}$$

where $A(\gamma J, \gamma^{\backslash} J^{\backslash})$ is the Einstein spontaneous emission transition probability rate, for a transition from the state. γJ to the state $\gamma^{\backslash} J^{\backslash}$.

The sum is over all state $\gamma^{\backslash} J^{\backslash}$ with $E(\gamma^{\backslash} J^{\backslash}) < E(\gamma J)$.

The Einstein probability rate is related to gf with the following relation by [17]:

$$gA = \frac{8\pi^2 e^2 \sigma^2}{mc} gf. \tag{12}$$

Since the natural lifetime $\tau(\gamma J)$ is the inverse of transition probability, then:

$$\tau(\gamma J) = \left(\sum A(\gamma J, \gamma' J')\right)^{-1} \tag{13}$$

which is applicable to an isolated atom.

Interaction with matter or radiation will reduce the lifetime of any state.

3. COMPUTATION OF GAIN COEFFICIENT

The possibility of laser emission from plasma of ions of various members of Na like Co via electron collisional pumping, in the XUV and soft X-ray spectral regions is investigated at different plasma temperatures and plasma electron densities.

The reduced population densities are calculated by solving the coupled rate equations [18-21].

$$N_j\left[\sum_{i<j} A_{ji} + N_e\left(\sum_{i<j} C_{ji}^d + \sum_{i>j} C_{ji}^e\right)\right]$$
$$= N_e\left(\sum_{i<j} N_i C_{ij}^e + \sum_{i>j} N_i C_{ij}^d\right) + \sum_{i>j} N_i A_{ij} \tag{14}$$

where N_j is the population of level j, A_{ji} is the spontaneous decay rate from level j to level i, C_{ji}^e is the electron collisional excitation rate coefficient, and C_{ji}^d is the electron collisional de-excitation rate coefficient, which is related to electron collisional excitation rate coefficient by [22-23].

$$C_{ji}^{d} = C_{ij}^{e}\left[\frac{g_i}{g_j}\right]\exp\left[\Delta E_{ji}/KT_e\right] \tag{15}$$

where g_i and g_j are the statistical weights of lower and upper levele, respectively.

The population of the j [th] level is obtained from the identity [19,20,24],

$$N_i = \frac{\left[N_j\right]\left[N_i\right]\left[N_t\right]}{N_i \quad N_t \quad N_e} N_e \tag{16}$$

where N_I is the total number density of all levels of the ion under consideration, and N_t is the total number density of all ionization stage.

Since the populations calculated from Equation (14) are normalized such that [19,20,25]

$$\sum_{J=1}^{55}\left(\frac{N_J}{N_I}\right) = 1 \tag{17}$$

where n is the number of all the levels of the ion under consideration.

Electron collisional pumping has been applied. Collision in the lasant ion plasma will transfer the pumped quanta to other levels, and resulted in population inversions then produced between the upper and lower levels.

Once a population inversion has ensured a positive gain through F > 0 [1].

$$F = \frac{g_u}{N_u}\left[\frac{N_u}{g_u} - \frac{N_l}{g_l}\right] \tag{18}$$

where $\frac{N_u}{g_u}$ and $\frac{N_l}{g_l}$ are the reduced populations of the upper level and lower level respectively. Equation (18) has been used to calculate the gain coefficient for Doppler broadening of the various transitions in the Na like Co ion.

$$\alpha = \frac{\lambda_{lu}^3}{8\pi}\left(\frac{M}{2\pi KT_i}\right)^{1/2} A_{ul} N_u F \quad (19)$$

where M is the ion mass, λ_{1u} is the transition wavelength in cm, T_i is the ion temperature in $^{\mathrm{o}}\mathrm{K}$ and u,l represent the upper and lower transition levels respectively.

The gain coefficient is expressed in terms of the upper state density (N_u). This quantity depends on how the upper state is populated, as well as on the density of the initial source state. The source state is often the ground state for a particular ion.

4. RESULTS AND DISCUSSIONS

4.1. Energy Levels

Adopting the program COWAN [17], we have computed the parameters of atomic structures of Co XVII. The energy levels considered in the calculation have 21 fine structures ranging from ground state $1s^2\ 2s^2\ 2p^6$ 3s to the excited states$1s^2\ 2s^2\ 2p^6$ 3d, $1s^2\ 2s^2\ 2p^6$ 4s, $1s^2\ 2s^2\ 2p^6$ 4d, $1s^2\ 2s^2\ 2p^6$ 5s, $1s^2\ 2s^2\ 2p^6$ 5s, $1s^2\ 2s^2\ 2p^6$ 5d, $1s^2\ 2s^2\ 2p^6$ 5g even parity levels and $1s^2\ 2s^2\ 2p^6$ 3p, $1s^2\ 2s^2\ 2p^6$ 4p, $1s^2\ 2s^2\ 2p^6$ 4f, $1s^2\ 2s^2\ 2p^6$ 5p, $1s^2\ 2s^2\ 2p^6$ 5f odd parity levels (Table 1).

Table 2 present energy levels and fine-structure splitting for Co^{16+} also presented the energy levels calculations of (Younis et al., 2006) the present calculations differ by less than 0.15% for most of the levels, they

Table 1. Calculated HFR energy levels and fine structure splitting.

(In 1000 cm^{-1}) for Co^{16+}					
index	Level	Term	$E_{cal.}$	$E_{[28]}$	$E_{[29]}$
1	$1s^2 2s^2 2p^6 3s$	$^2s_{1/2}$	0	0	0
2	$1s^2 2s^2 2p^6 3p$	$^2p_{1/2}$	296.434	294.556	294.537
3	$1s^2 2s^2 2p^6 3p$	$^2p_{3/2}$	320.889	322.404	319.94
4	$1s^2 2s^2 2p^6 3d$	$^2d_{3/2}$	722.116	719.577	720.211
5	$1s^2 2s^2 2p^6 3d$	$^2d_{5/2}$	726.32	724.36	723.915
6	$1s^2 2s^2 2p^6 4s$	$^2s_{1/2}$	2081.824	2079.55	2079.55
7	$1s^2 2s^2 2p^6 4p$	$^2p_{1/2}$	2199.799	2196.5	2196.5
8	$1s^2 2s^2 2p^6 4p$	$^2p_{3/2}$	2209.446	2211.204	2206.58
9	$1s^2 2s^2 2p^6 4d$	$^2d_{3/2}$	2357.002	2353.713	2353.69
10	$1s^2 2s^2 2p^6 4d$	$^2d_{5/2}$	2358.815	2355.236	2355.25
11	$1s^2 2s^2 2p^6 4f$	$^2f_{5/2}$	2421.119	2419.661	2419.79
12	$1s^2 2s^2 2p^6 4f$	$^2f_{7/2}$	2421.767	2420.361	2420.41
13	$1s^2 2s^2 2p^6 5s$	$^2s_{1/2}$	2971.567	2967.7	2967.7
14	$1s^2 2s^2 2p^6 5p$	$^2p_{1/2}$	3029.815	3026.1	3026.1
15	$1s^2 2s^2 2p^6 5p$	$^2p_{3/2}$	3034.58	3028.187	3030.8
16	$1s^2 2s^2 2p^6 5d$	$^2d_{3/2}$	3105.86	3102.321	3102.2
17	$1s^2 2s^2 2p^6 5d$	$^2d_{5/2}$	3106.794	3102.92	3103
18	$1s^2 2s^2 2p^6 5f$	$^2f_{5/2}$	3138.465	3135.506	3135.2
19	$1s^2 2s^2 2p^6 5f$	$^2f_{7/2}$	3138.798	3135.796	3135.3
20	$1s^2 2s^2 2p^6 5g$	$^2g_{7/2}$	3142.107	------	3139.5
21	$1s^2 2s^2 2p^6 5g$	$^2g_{9/2}$	3142.304	------	3139.5

Table 2. Wavelength, oscillator strength and radiative rate for allowed.

Transititions in Co^{17+}						
i	j	λ(Å)	$A_{ji}(sec)^{-1}$	$fij_{cal.}$	$fij_{Ref.1}$	$fij_{Ref.2}$
1	2	337.343	7.10E+09	1.21E−01	1.18E−01	0.119
1	3	311.634	1.80E+10	2.62E−01	2.60E−01	0.262
1	7	45.459	2.43E+11	7.52E−02	1.88E−01	0.07862
1	8	45.26	4.92E+11	1.51E−01	3.75E−01	0.1459
1	14	33.005	1.41E+11	2.30E−02	1.69E−02	0.02456
1	15	32.953	2.83E+11	4.60E−02	3.37E−02	0.05223
2	4	234.917	1.69E+10	2.80E−01	2.97E−01	0.278
2	6	56.01	1.36E+11	6.41E−02	8.68E−02	0.1788
2	9	48.53	3.45E+11	3.24E−01	2.16E−01	0.3112
2	13	37.381	6.00E+10	1.26E−02	1.03E−02	0.0122
2	16	35.594	2.65E+11	1.01E−01	1.47E−01	0.1027
3	4	249.235	2.83E+09	2.64E−02	2.79E−02	0.026
3	5	246.651	1.75E+10	2.40E−01	2.54E−01	0.237
3	6	56.788	2.62E+11	6.32E−02	9.49E−02	0.092
3	9	49.113	8.88E+10	3.21E−02	2.35E−02	0.0326
3	10	49.069	0.5333+12	0.289028	0.2111	0.2932
3	13	37.726	1.17E+11	1.25E−02	1.08E−02	0.01176
3	16	35.907	5.15E+10	9.97E−03	1.58E−02	0.00988
3	17	35.895	3.10E+11	8.97E−02	1.42E−01	0.09001

4	7	67.674	4.58E+10	3.13E−02	7.83E−02	0.0303
4	8	67.235	9.33E+09	6.31E−03	1.58E−02	0.0062
4	11	58.858	1.80E+12	9.30E−01	7.63E−01	0.93
4	14	43.333	1.81E+10	5.09E−03	1.47E−04	0.00508
4	15	43.244	3.65E+09	1.02E−03	2.94E−05	0.001
4	18	41.385	6.63E+11	1.70E−01	3.07E−01	0.1696
5	8	67.425	5.55E+10	3.77E−02	9.55E−02	0.0368
5	11	59.004	8.48E+10	4.42E−02	3.65E−02	0.0443
5	12	58.982	1.70E+12	8.85E−01	7.29E−01	0.9
5	15	43.323	2.17E+10	6.12E−03	2.20E−04	0.00601
5	18	41.457	3.13E+10	8.07E−03	1.47E−02	0.00825
5	19	41.451	6.27E+11	1.61E−01	2.94E−01	0.1625
6	7	847.636	1.62E+09	1.75E−01	1.60E−01	0.16
6	8	783.565	4.10E+09	3.77E−01	3.59E−01	0.37
6	14	105.486	4.78E+10	7.98E−02	2.53E−02	0.0853
6	15	104.959	9.70E+10	1.60E−01	5.07E−02	0.169
7	9	636.12	3.65E+09	4.45E−01	4.28E−01	0.41
7	13	129.768	4.32E+10	1.09E−01	7.17E−02	0.103
7	16	110.368	7.70E+10	2.81E−01	5.46E−02	0.289
8	9	677.707	6.05E+08	4.17E−02	3.88E−02	0.039
8	10	669.479	3.77E+09	3.80E−01	3.53E−01	0.35
8	13	131.213	8.30E+10	1.07E−01	7.65E−02	0.102
8	16	111.556	1.49E+10	2.78E−02	4.55E−03	0.028
8	17	111.439	8.98E+10	2.51E−01	4.10E−02	0.257
9	11	1559.639	2.90E+08	1.06E−01	1.34E−01	———
9	14	148.63	2.15E+10	7.11E−02	6.27E−02	0.0713
9	15	147.585	4.38E+09	1.43E−02	1.26E−02	0.014
9	18	127.965	2.98E+11	7.33E−01	6.10E−02	0.699
10	11	1605.036	1.27E+07	4.88E−03	6.21E−03	———
10	12	1588.536	2.62E+08	9.88E−02	1.26E−01	———
10	15	147.981	2.62E+10	8.57E−02	7.67E−02	0.0845
10	18	128.263	1.41E+10	3.48E−02	2.97E−03	0.0345
10	19	128.208	2.83E+11	6.96E−01	5.94E−02	0.7218
11	16	146.041	8.05E+09	1.72E−02	1.25E−01	0.017
11	17	145.842	3.85E+08	1.23E−03	8.91E−03	0.0013
11	20	138.699	3.48E+11	1.34E+00	———	———
12	17	145.98	7.68E+09	1.84E−02	1.33E−01	0.018
12	20	138.823	1.29E+10	3.71E−02	———	———
12	21	138.785	3.60E+11	1.30E+00	———	———
13	14	1716.791	5.10E+08	2.26E−01	2.18E−01	———
13	15	1586.972	1.30E+09	1.26E−02	4.52E−01	———
14	16	1315.008	1.14E+09	5.92E−01	5.54E−01	———
15	16	1402.913	1.88E+08	5.55E−02	5.38E−02	———
15	17	1384.773	1.17E+09	5.06E−01	4.89E−01	———
16	18	3067.075	1.34E+08	1.90E−01	1.68E−01	———
17	18	3157.503	5.87E+06	8.77E−03	7.87E−03	———
17	19	3124.561	1.21E+08	1.77E−01	1.59E−01	———
18	20	27454.79	7.39E+04	1.11E−02	———	———
19	20	30225.63	2.05E+03	2.80E−04	———	———
19	21	28522.53	6.83E+04	1.04E−02	———	———

have used Configuration–Interaction Code (CIV3).and the differ from (Nist, 2009) by less than 0.15% for most of the levels, except in some cases like the levels 2, 3, 4 and 5 in which the percentage difference between our value and the last two calculations is (0.63%, 0.64%), (−46%, 29%), (0.35%, 0.26%) and (0.27%, 0.33%) respectively, This mean that our results are in a good agreement with the theoretical and experimental value.

4.2. Oscillator Strength and Transition Probability

In the present work, we report results of oscillator strength and transition probabilities using the Cowan code taking relativistic corrections into account.

Table 3 present the wavelength, transition probability and the values of oscillator strength with a comparison against another published paper (Younis et al., 2006 [Reff.1]) and NIST. There was good agreement with (NIST, 2009 [Reff.2]) and only small differences about (0.8%) there was some similarities with the paper and high differences in some values about (85%).

4.3. Radiative Lifetime

Table 4 contains the present results of radiative lifetime for the upper and lower laser levels for the sodium-like Co.

The present calculations predict that the lifetime of the upper laser level must be longer than the lifetime of lower one to ensure the fast depletion of the population of lower level which helps the laser to sustain continuous or quasicontinuous wave operation.

4.4. Gain Distributions

4.4.1. Level Populations

The reduced population densities are calculated for 4 levels by solving the coupled rate Equations (14) using the coupled equations CRMO code [26] for solving simultaneous coupled rate equations.

Our calculations for the reduced population as a function of electron densities are plotted in Figures 1 and 2 at three different plasma temperatures (1/2, 3/4 of the ionization potential) for Na like Co.

We took into account in the calculation spontaneous radiative decay rate and electron collisional processes between all levels under study.

The behavior of level populations of the various ions can be explained as follows: in general, at low electron densities the reduced population density is proportional to the electron density, where excitation to an excited state is followed immediately by radiation decay, and collisional mixing of excited levels can be ignored.

At high densities ($N_e > 10^{20}$), radiative decay to all levels will be negligible compared to collisional depopulations and all level population become independent of electron density and approximately equal (see Figures 1 and 2). The population inversion is largest where electron collisional de-excitation rate for the upper level is comparable to radiative decay for this level [21].

Table 3. Radiative lifetime for Co^{16+} laser levels.

Index	Level	life time (ns)
6	$4s(^2s_{1/2})$	2.51E−03
7	$4p(^2p_{1/2})$	3.44E−03
8	$4p(^2p_{3/2})$	1.78E−03
9	$4d(^2d_{3/2})$	2.28E−03

Table 4. Parameters of the most intense laser transitions in Co^{16+} ion plasma.

Transition	Atomic data Ion	Co XVII
$4p\ (^2p_{1/2}) \rightarrow 4s\ (^2s_{1/2})$	Wavelength λ (Å)	847.6
	Maximum gain α (cm^{-1})	0.162
	Electron density N_e (cm^{-3})	3.50E+19
	Electron temperature T_e (eV)	409.9
$4d\ (^2d_{3/2}) \rightarrow 4p\ (^2p_{3/2})$	Wavelength λ (Å)	677.7
	Maximum gain α (cm^{-1})	0.128
	Electron density N_e (cm^{-3})	5.00E+19
	Electron temperature T_e (eV)	409.9
$5p\ (^2p_{1/2}) \rightarrow 5s\ (^2s_{1/2})$	Wavelength λ (Å)	....
	Maximum gain α (cm^{-1})	....
	Electron density N_e (cm^{-3})	....
	Electron temperature T_e (eV)	....
$5d\ (^2d_{3/2}) \rightarrow 5p\ (^2p_{3/2})$	Wavelength λ (Å)	....
	Maximum gain α (cm^{-1})	....
	Electron density N_e (cm^{-3})	....
	Electron temperature T_e (eV)	
$5d\ (^2d_{5/2}) \rightarrow 5p\ (^2p_{3/2})$	Wavelength λ (Å)	....
	Maximum gain α (cm^{-1})	....
	Electron density N_e (cm^{-3})	....
	Electron temperature T_e (eV)	....

Form our study, it was found that the gain coefficient was very low at (1/4 ionization potential) for all elements, so the results of gain coefficient and reduced population have not been mentioned.

4.4.2. Inversion Factor

As we mentioned before, laser emission will occur only if there is population inversion or in other words for positive inversion factor F >

0. In order to work in the XUV and X-ray spectral regions, we have chosen transitions between any two levels producing photons with wavelength between 30 and 1000 Å.The electron density at which the population reaches collisional equilibrium

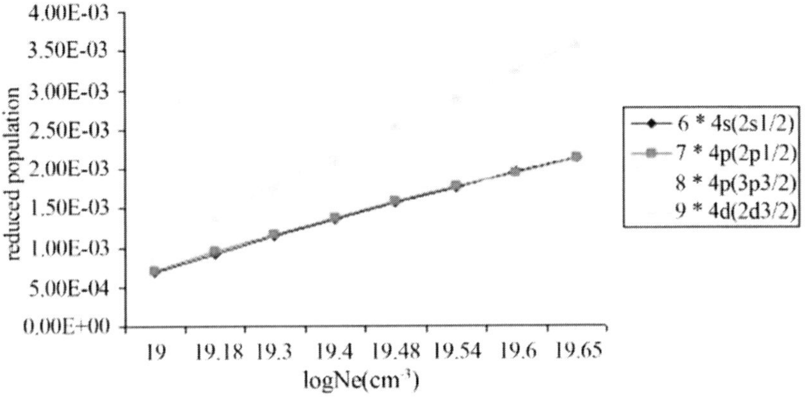

Figure 1. Reduced population of Co^{16+} levels after electron collisional pumping as a function of the electron density at temperature 273.29 eV.

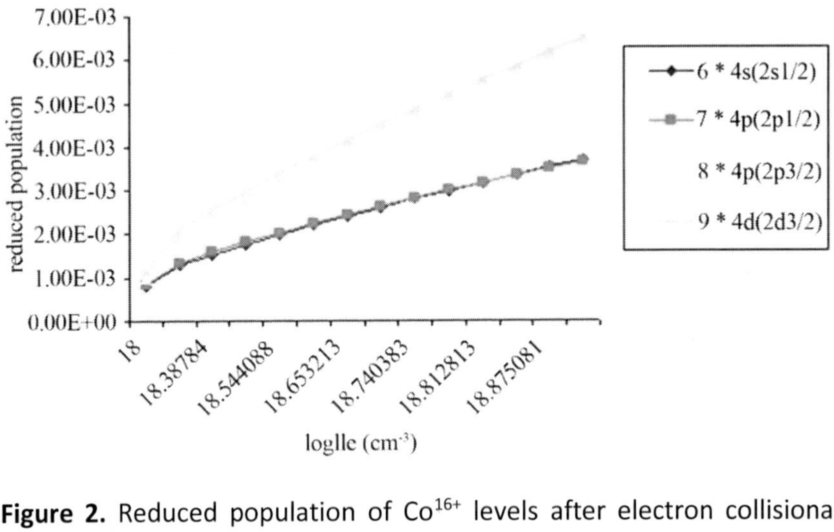

Figure 2. Reduced population of Co^{16+} levels after electron collisional pumping as a function of the electron density at temperature 409.94 eV.

approximately equal to A/D, where A is radiative decay rate and D is the collisional de-excitation rate [20]. The population inversion is largest where the electron collisional de-excitation rate for the upper level is comparable to the radiative decay rate for this level.

For increasing atomic number Z, the population inversion occurs at higher electron densities, this is due to the increase in the radiative decay rate with Z and the decrease in collisional de-excitation rate coefficient with Z [27].

4.4.3. Gain Coefficient

As results of population inversion there will be positive gain in laser medium. Equation (19) has been used calculate gain coefficient for the Doppler broadening of various transitions in the Na like Co.

Our results for the maximum gain coefficient in cm^{-1} for those transitions having a positive inversion factor F > 0 in the case of Co^{16+} ion at different temperatures are calculated and plotted against electron density in Figures 3 and 4. The figures show that the population inversions occur for several transitions in the Co^{16+} ion, however the largest gain occurs for the Co^{16+} ion at $4d(^2p_{1/2}) - 4\ s(^2s_{1/2})$ transition.

For Na-like Co, the population inversion is due to strong monopole excitation from the 3s ground state to the 3s 4d configuration and also the radiative decay of the 3s 4d level to the ground level is forbidden, while the 3s 4p level decays very rapidly to the ground level.

This short wavelength laser transitions was produced using plasmas created by optical lasers as the lasing medium. It is obvious that the gain increases with the temperature as the maximum gain increases with atomic number. Moreover, the peak of the gain curves shifts to higher electron densities with the increase of atomic number.

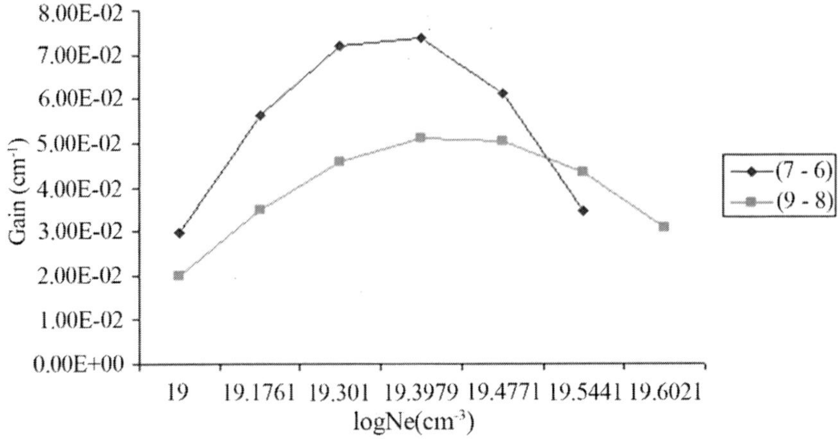

Figure 3. Gain coefficient of possible laser transitions against electron density at temperature 273.29 eV in Co^{16+}.

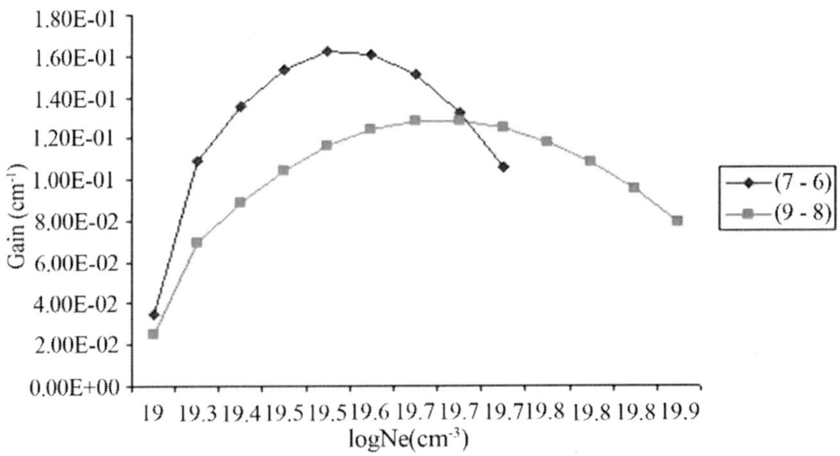

Figure 4. Gain coefficient of possible laser transitions against electron density at temperature 409.94 eV in Co^{16+}.

5. CONCLUSION

The analysis that has been presented in this work shows that electron collisional pumping (ECP) is suitable for attaining population inversion and offering the potential for laser emission in the spectral region between 50 and 1000 Å from the Na like Co. This class of lasers can be achieved under the suitable conditions of pumping power as well as electron density. If the positive gains obtained previously for some transitions in (Co^{16+} ion), together with the calculated parameters could be achieved experimentally, then successful low cost electron collisional pumping XUV and soft X-ray lasers can be developed for various applications. The results have suggested the following laser transitions in the Co^{16+} plasma ion, as the most promising laser emission lines in the XUV and soft X-ray spectral regions.

REFERENCES

1. R. C. Elton, "X-Ray Lasers," Academic Press, New York, INC, 1990.

2. J. E. Trebes, S. B. Brown, et al., "Demonstration of XRay Holography with an X-Ray Laser," Science, Vol. 238, No. 4826, 1987, pp. 517-519. . B. Da Silva, J. E. Trebes, et al., "X-Ray Laser Microscopy of Rat Sperm Nuclei," Science, Vol. 258, No. 5080, 1992, pp. 269-271.

3. D. Ress, L. B. Da Silva, et al., "Measurement of LaserPlasma Electron Density with a Soft X-Ray Laser Deflectometer," Science, Vol. 265, v5171, 1994, pp. 514-517.

4. R. Cauble, L. B. Da Silva, et al., "Simultaneous Measurement of Local Gain and Electron Density in X-Ray Lasers," Science, Vol. 273, No. 5278, 1996, pp. 1093-1096.

5. D. H. Kalantar, M. H. Key, L. B. Da Silva, et al., "Measurement of 0.35 μm Laser Imprint in a Thin Si Foil Using an X-Ray Laser

Backlighter," Physical Review Letters, Vol. 76, 1996, 19, pp. 3574-3577.

6. C. F. Fischer, "Method: MCHF (ab Initio)," 2002. http://www.Vuse.Vanderbilt.Edu/~cff /mchfcollection/

7. W. Siegel, J. Migdalek and Y. K. Kim, "Dirac-Fock Oscillator Strengths for E1 Transitions in the Sodium Isoelectronic Sequence (Na I-Ca X)," Atomic Data and Nuclear Data Tables, Vol. 68, No. 2, 1998, pp. 303-322.

8. D. A. Verner, P. D Barthel and D. Tyther, "Atomic Data for Absorption Lines from the Ground Level at Wavelengths Greater than 228 A," Astronomy and Astrophysics Supplement Series, Vol. 108, No. 2, 1994, pp. 287-340.

9. J. R. Fuhr and W. L. Wiese, "Atomic Transition Probabilities," In: D. R. Lide, Ed., CRC Handbook of Chemistry and Physics, 72nd Edition, CRC Press, 1991, pp. 211-213.

10. W. L. Wiese, M. W. Smith and B. M. Miles, "Atomic Transition Probabilities (Na through Ca-A Critical Data Compilation)," National Standard Reference Data Series, National Bureau of Standards (US), NSRDS-NBS 22, Vol. II, 1969.

11. G. A. Martin, J. R. Fuhr and W. L. Wiese, "Atomic Transition Probabilities-Scandium: Through Manganese," Journal of Physical and Chemical Reference Data, Vol. 17, Suppl. 3, 1988,

12. J. R. Fuhr, G. A. Martin and W. L. Wiese, "Atomic Transition Probabilities-Iron: Through Manganese," Journal of Physical and Chemical Reference Data, Vol. 17, Suppl. 4, 1988, pp. 167-171.

13. R. M. Lowe and E. Biemont, "Lifetime measurements for the 4p 2Po and 5p 3Po Levels and Calculation of Transition Probabilities in Na I," Journal of Physics b-Atomic Molecular and Optical Physics, Vol. 27, No. 11, 1994, pp. 2161-2167.

14. D. R. Hartree and E. E. Salpeter, "Quantum Mechanics of One- and Two-Electron Atoms," Springer-Verlage, Berlin and New York, 1657.

15. I. Sobelman, "Atomic Spectra and Radiative Transition," Springer, Berlin, 1979.

16. R. D. Cowan, "The Theory of Atomic Structure and Spectra," University of California Press, Berkeley, 1981.

17. U. Feldman, A. K. Bhatia and S. Suckewer, "Short Wavelength Laser Calculations for Electron Pumping in NeonLike Krypton (Kr XXVII)," Journal of Applied Physics, Vol. 54, No. 5, 1983, pp. 2188-2197.

18. U. Feldman, J. F. Seely and G. A. Doschek, "3s-3p Laser Gain and x-Ray Line Ratios for the Carbon Isoelectronic Sequence," Journal of Applied Physics, Vol. 59, No. 12, 1986, pp. 3953-3957.

19. U. Feldman, G. A. Doschek, J. F. Seely and A. K. Bhatia, "Short Wavelength Laser Calculations for Electron Pumping in BE I and B I Isoelectronic Sequences (18 ≤ Z ≤ 36)," Journal of Applied Physics, Vol. 58, No. 8, 1985, p. 2909.

20. U. Feldman, J. F. Seely and A. K. Bhatia, "Scaling of Collisionally Pumped 3s-3p Lasers in the Neon Isoelectronic Sequence," Journal of Applied Physics, Vol. 56, No. 9, 1984, pp. 2475-2478.

21. G. Chapline and L. Wood, "X-Ray Lasers," Physics Today, Vol. 28, No. 40, 1975.

22. A. V. Vinogradov and V. N. Shlyaptsev, "Calculations of Population Inversion Due to Transitions in Multiply Charged Neon-Like Ions in the 200âEUR"2000 Ã... Range," Soviet Journal of Quantum Electronics, Vol. 10, No. 6, 1980, p. 754.

23. U. Feldman, J. F. Seely and G. A. Doschek, "Short Wavelength Laser Calculations in the Be I, B I And C I Isoelectronic

Sequences," Journal de Physique Collogues, Vol. 47, No. C6, 1986, p. C6-187.

24. M. J. Seaton, "Atomic Data for Opacity Calculations. I. General Description," Journal of Physics B: Atomic and Molecular Physics, Vol. 20, No. 23, 1987, 6363.

25. S. H. Allam, "CRMO Computer Code," Private Communication, 2003.

26. U. Feldman, J. F. Seely and A. K. Bhatia, "Density Sensitive x-Ray Line Ratios in the BE i, B i, and Ne i Isoelectronic Sequences," Journal of Applied Physics, Vol. 58, No. 11, 1984, pp. 3954-3958.

27. W. O. Younis, S. H. Allam and Th. M. El-Sherbini, "Fine-Structure Calculations of Energy Levels, Oscillator Strengths, and Transition Probabilities for Sodium-Like Ions (Co XVII-Kr XXVI)," Atomic Data and Nuclear Tables, Vol. 92, No. 2, 2005, pp. 187-205.

28. "Atomic Spectra Database Lines Form," NIST, Physical Meats Laboratory, Vol. 11, 2009, pp. 3954-3958.

Chapter 3

Optical Coherence in Astrophysics: The Powerful Alternative of Big Bang

Jacques Moret-Bailly

Physics Department, University of Burgundy, Dijon, France

ABSTRACT

The coherence of the interaction of light with a collisionless gas (Einstein 1917) founds the theory of gas lasers. It is, for the understanding of universe, a simpler and more powerful tool than the big bang which requires questionable supplements (dark matter, MOND, etc.). The Impulsive Stimulated Raman Scattering (ISRS) redshifts gradually light pulses which cross excited atomic hydrogen H*, so that the redshift is a measure of the column density of H*. Thus, the distance of the hot stars, surrounded by much H*, is exaggerated by the use of Hubble's law. Local exaggerated distances create voids in the maps of galaxies which become spongy. The interpretation of spectra of quasars, the periodicity of galaxy redshifts introduce an experimental "Karlsson's constant" exactly computed by ISRS. The need for dark matter comes from the exaggeration of the distance, therefore the size of galaxies. Without dark matter, celestial mechanics provides a reliable distance of spiral galaxies. Coherence also introduces superradiance and mode competition that explain that only the limbs of Strömgren spheres are visible as circles maybe punctuated by an even number of dots: Too numerous, the figures assigned to gravitational lenses can be such limbs. The coincidence of the ignition of the rings of SNR1987A with the extinction of the star is due to a multiphoton coherent scattering of star light, which amplifies the superradiant emission of the rings. A blueshift of microwaves crossing H* resulting, between 10 and 15 AU, of the expansion of solar wind, explains the "anomalous acceleration" of Pioneer probes. All is obtained without any change in theories of standard spectroscopy.

Keywords: Coherent Interactions of Light with Gas and Plasma; Emission; Absorption; Scattering; Nonlinear Phenomena; Astrophysics

1. INTRODUCTION

In a paper in Reviews of Modern Physics entitled "Laser physics: Quantum controversy in action." [1] W. E. Lamb Jr., W. P. Schleich, M. O. Scully and C. H. Townes criticize an abusive use of the word photon. For them, in accordance with the principles of quantum electrodynamics, the photon is a pseudo-particle resulting from the quantization of normal modes of the electromagnetic field. These modes should not only be defined, but also set by a physical system such as a cavity or a pulse propagating in a diffraction limited beam.

A diffraction limited beam may be obtained putting a diaphragm on a wave surface of a monochromatic plane wave. The diaphragm is the waist of the beam which, including the virtual side, expands in the past and the future. As the transverse extension of a pulse is very large at a long distance, the field defining the mode may excite atoms in a large volume. In quantum theory, a large set of identical atoms may be considered as a single, highly degenerated system which interacts with the pulse, then, by decoherence, transfers a quantum to a single atom. In classical theory, all atoms are slightly excited, then decoherence results from an electromagnetic transfer of energy to an atom favored by a fluctuation of the zero point field.

On the contrary, some physicists believe that the photon is a small, real particle:

In 1931, Menzel [2] wrote: It is easily proved that the so-called "stimulated emissions" are unimportant in the nebulae. Menzel's demonstration is doubtful because in this paper, he confuses two photometric quantities, radiance and irradiance: The radiance measures the energy flow in a light beam, it is constant in a transparent medium, while the irradiance is the flow, coming from all directions, which crosses from a given side, a surface of unit area. This confusion was common at this time because these two quantities differ by a solid angle often not taken into account in the equations of dimension. Menzel's opinion remains in astrophysics: For most astrophysicists, the small photon in the extremely low pressure gas of a nebula can interact only with a single molecule. This interaction has a stochastic nature requiring, for instance the use of a Monte-Carlo draw [3].

Without taking sides in this controversy, it is useful to precise which theory of the propagation of light is used. Here, we follow the choice done by Lamb et al.: The interaction of an electromagnetic wave with a molecule follows to a deterministic process, so that Huygens' construction may apply and the interaction is coherent.

The view of Lamb et al. is supported experimentally by the absence of blue sky in the stratosphere and the need to avoid incoherent diffusions in gas lasers: All light interactions with low pressure gas are coherent. Incoherences generating the blue sky, result of collisions, mostly binary, whose density is proportional to the square of the pressure.

To simplify, in our models, nebulae consist of pure hydrogen.

Section 2 summarizes the general properties of coherent spectroscopy.

Section 3 studies, using coherent spectroscopy, theoretical models involving extremely hot sources, huge systems, which may exist in astronomy.

Section 4 compares some observations with these models.

2. COHERENT INTERACTIONS

2.1. Huygens' Construction

Huygens' constructions for any sine waves are representations of the generation of a new wave surface near an old wave surface: Sources located in the vicinity of the old surface wave radiate synchronous spherical waves. The envelope of the spheres obtained after a time Δt is a new wavefront. The part of the envelope corresponding to a backward wave is eliminated because a wavefront having an initial offset of a quarter wavelength, generates back a wave of opposite phase which destroys the backward wave.

Figure 1(a) is Huygens' construction in an homogeneous medium. The volume between two wave surfaces whose distance is infinitesimal is divided into infinitesimal parcels which radiate in phase to build the new wavefront.

Figure 1(b), on the diffusion of an electromagnetic wave, adds, to diffusion 1A, a diffusion by a large number of molecules. In a transparent medium, this last release is delayed by $\pi/2$, so that the sum of two sine waves is a sine wave slightly delayed, called refracted.

Figure 1(c) shows that the wavelets emitted by rare molecules cannot generate an envelope: it is an inco herent scattering. In gases, the collisions of two molecules create binary systems depending on many parameters, so it creates different molecular systems that produce incoherent diffusions.

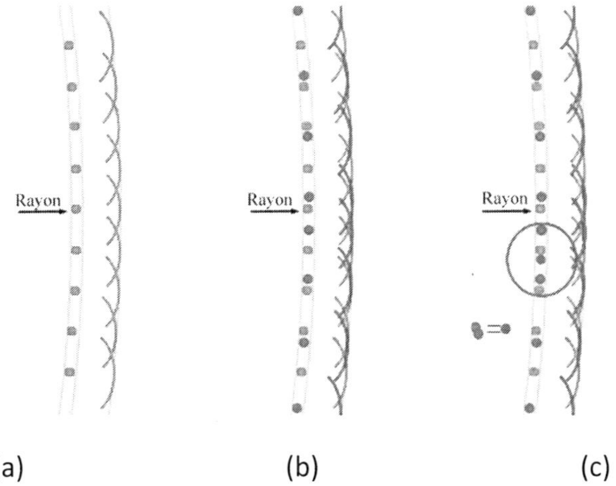

Figure 1. Huygens' constructions: (a) in a continuous medium; (b) in a low pressure gas; (c) including rare molecules possibly made up by two colliding molecules.

Similarly, a gas laser can only operate if the incoherent diffusions are quite unlikely. To avoid a loss of too much energy, the pressure is about 100 Pa.

2.2. Superradiance

When a light beam propagates in a gas, its spectral radiance $I(\nu)$ varies according to the law given by Einstein in 1917 [4]

$$\mathrm{d}I(\nu) = B(\nu)I(\nu)\mathrm{d}x \tag{1}$$

where $\mathrm{d}x$ is the path in the gas and $B(\nu)$ Einstein's coefficient which depends on the nature and state of the gas. (We generally imply "(ν)" and "spectral", which specify that it is a flow of energy within a range of frequencies $\mathrm{d}\nu$ or wavelengths of $\mathrm{d}\lambda$). The integration of Equation (1) introduces a constant we write I_0 assuming it is the absolute radiance of a ray from a blackbody at 0 K, so that $I_0 = h\nu^3/c^2$ on average[1] (Planck 1911 [5]). If B is constant, I is an exponential function of x path. It is often appropriate to say that there is a superradiance when I is greater than $2I_0$, while I usually remains close to I_0.

2.3. Competition of Modes

Amplified in a volume G, (Figure 2) the rays absorb energy, so that coefficient B decreases. Formula 1 shows that the strongest ray absorbs more energy than the weaker, the radiance of which remains close to I_0: It remains bright only the rays which would be the strongest taking no account of superradiance.

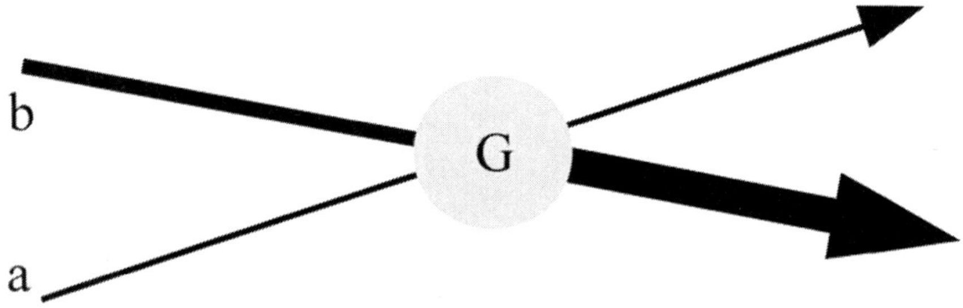

Figure 2. Competition of modes.

2.4. Multiphotonic Interactions

Suppose that an extremely hot source emits a continuous spectrum. At each frequency, its high radiance beams can pump atoms to virtual levels, from which they may be pumped, and other pumpings may allow them reach a stationary high level: by combinations of frequencies, the whole continuous spectrum is involved in the pumping of the atoms to stationary levels.

If the decay from this stationary level feeds a superradiant beam, the B coefficient may remain high, so that there is a transfer of energy from the whole continuous spectrum of the source to the superradiant beam.

This transfer is only possible if it increases the entropy of the system. The temperature of a light ray at a frequency v is deduced by Planck's law from its spectral radiance and v. To increase the entropy, the temperature of the rays emitted by the source and the temperature of the superradiant beam tend to a common limit: A fraction of the energy of the continuous spectrum emitted by the source is transferred to the superradiant beam.

2.5. Impulsive Stimulated Raman Scattering (ISRS)

This scattering is widely used in Keith Nelson's laboratory in MIT, mainly to study matter. A simple description may be found in paper [6], for a review of ISRS see, for instance, Dhar et al. [7] or Lee et al. [8].

The ISRS may be considered as a supplement of refraction obtained by an addition of a Raman coherent scattering to the Rayleigh coherent scattering which produces the refraction (Look at appendix). Just as the Rayleigh coefficient of scattering, the Raman coefficient of scattering deduced from the quadrupolar polarizability is, in first approximation, far from the resonances, proportional to the frequency of the incident, exciting ray.

However, in refraction, the incident and Rayleigh scattered beams have the same frequency, so that they add into a wave at the same frequency. In ISRS, the exciting and coherently scattered beams have different frequencies, beats appear.

The fields scattered at two points A and B of a ray whose distance is x, have the same phase-shifts versus the incident wave. Along the path AB, they get the phase-shift $\Delta\phi = 2\pi x(1/\lambda_i - 1/\lambda_d)$, where λ_i and λ_d are the wavelengths of the incident and scattered waves. λ_i and λ_d differ except for particular rays propagating in some crystals. Thus, assuming a negligible variation of the intensity of the incident beam, for a distance x such that $\Delta\phi$ equals π, the scattered amplitudes cancel.

Figure 3 shows, with unrealistic orders of magnitude, interference of an incident wave with an other whose frequency is slightly different and amplitude lower, playing the role of a Raman Stokes wave. The amplitude of the sum of both waves decays until their phases are opposed. This sum has the appearance of a sine wave whose frequency decreases then increases.

The decay zone is selected by stopping the interaction by the use of pulses "shorter than all involved time constant" (G. L. Lamb [9]). If this condition is met, the beating of incident and scattered waves does not appear and their sum has a sinusoidal appearance. Fourier analysis shows that the major frequency is intermediate between initial frequencies in the ratio of the amplitudes of the components.

The computation of the frequency shift in the appendix shows that:

- The properties of the ISRS that do not involve temporal variations throughout the duration of the pulse are obtained by a simple transposition of the properties of refraction.
- Refraction and ISRS are spatially coherent, allowing the formation of images.
- Index of refraction, and relative variation of frequency by ISRS $\Delta f / f$ depend on a dispersion which is low except in the vicinity of resonances.

The ISRS frequency shift is easily observed by using femtosecond laser pulses [6].

The ISRS excites the active medium during the pulses. The decay of this excitation can be probed by another ISRS. It can also be probed with a continuous laser beam which undergoes a frequency variation during the pulsed excitation of the Raman level.

A coupling of several ISRS so that the balance of excitations of the medium equals zero is called “parametric”. The matter is a catalyst and the frequency shifts are much more intense because the Raman transitions are not saturated. Evidently, energy exchanges must be in accordance with the principles of thermodynamics.

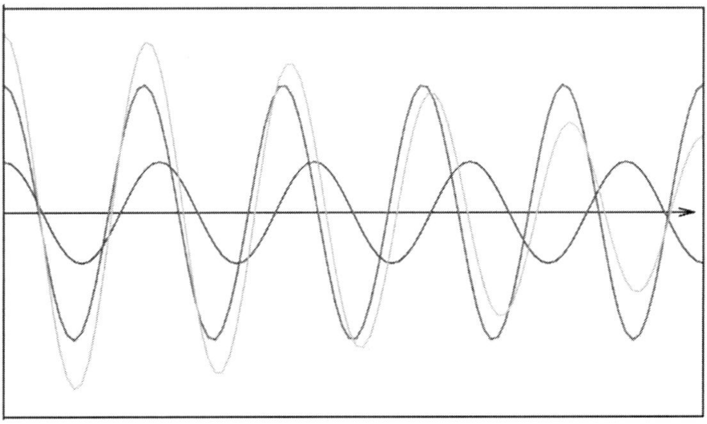

Figure 3. Addition of sine wave of different frequencies: While two sine curves * * have a constant amplitude, their sum* shows the beginning of a beat.

2.6. Coherent Raman Effect on Incoherent Light (CREIL)

CREIL is an ISRS wherein the pulses result from the usual representation of the temporally incoherent light emitted by thermal sources. The 10 femtosecond pulses are replaced by nanosecond pulses. The appendix shows that an increase of the length of the pulses by a factor k decreases the order of magnitude of ISRS by a factor $k^3 = 10^{15}$, so that an observation of CREIL requires an astronomical path of interaction.

A thermal electromagnetic field is always present, always tending to lower the frequency of the light.

The only gas, abundant in the universe, having a resonance period of more than, and close to 1 ns, thus able to catalyze a CREIL, is excited atomic hydrogen in the levels n = 2, denoted H*: it has quadrupole transitions at 178 MHz in the state $2S_{1/2}$, 59 Mhz in $2P_{1/2}$ and 24 MHz in $2P_{3/2}$. More excited levels have resonances of lower frequencies, their effect being almost negligible.

3. ASTRONOMICAL MODELS

3.1. Strömgren's Sphere [10]

In 1939, Strömgren [10] has defined a system consisting of a source (star) very hot (several hundreds of thousands of kelvins) placed in a cloud of hydrogen, at low pressure, initially cold.

The source generates a sphere of plasma almost completely ionized into protons and electrons. Strömgren showed that some atoms created by the cooling of the plasma radiate much energy, which accelerates the cooling and creates new atoms in a catastrophic process. It therefore appears a shell of plasma containing hydrogen atoms that radiate their spectral lines intensely. Strömgren seems to have ignored the superradiance, so that his system had the appearance of Figure 4.

The shell is divided into an infinite number of sublayers (Figure 5). As a ray intersects the sphere, the path in each sub-layer is an increasing function of the distance r of the ray to the star. So radiance I is a function of r, first increasing, finally falling to I_0, so it has at least a maximum R which specifies the radius of the Strömgren's sphere. Accordingly, only the rays at the limb of the sphere are brilliant.

If superradiance is sufficient the symmetry is broken by competition of modes in the limb. The limb is seen as an even number of points because the phases of two adjacent modes are opposite (Figure 6).

Figure 4. Strömgren's shell without superradiance.

3.2. Emission inside a Strömgren's Sphere

The sphere contains mostly free protons and electrons that do not interact with light. A low density of excited atoms increases nearly exponentially in terms of r, to the surface.

Suppose impurities absorb enough frequencies around Lyman alpha inside the sphere, so that the radiance of the rays which arrive from inside in vicinity of the surface is $h\nu^3/c^2$. Study light in the vicinity of the exit point of the sphere :

- In depth, the emission (amplification) of a line, for example Lyα at λ_0 is small, then it is redshifted by

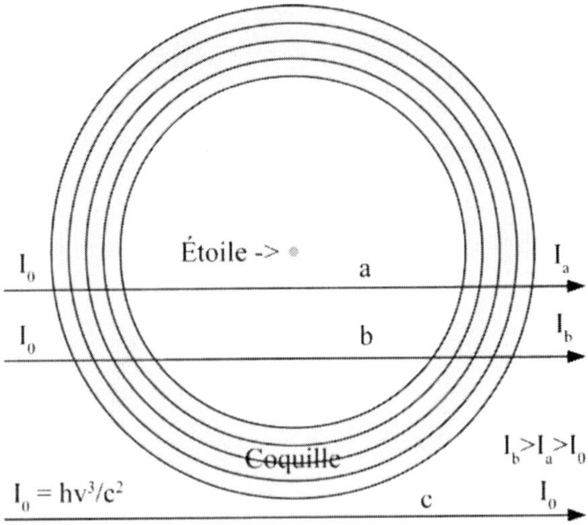

Figure 5. Emission of the shell (coquille).

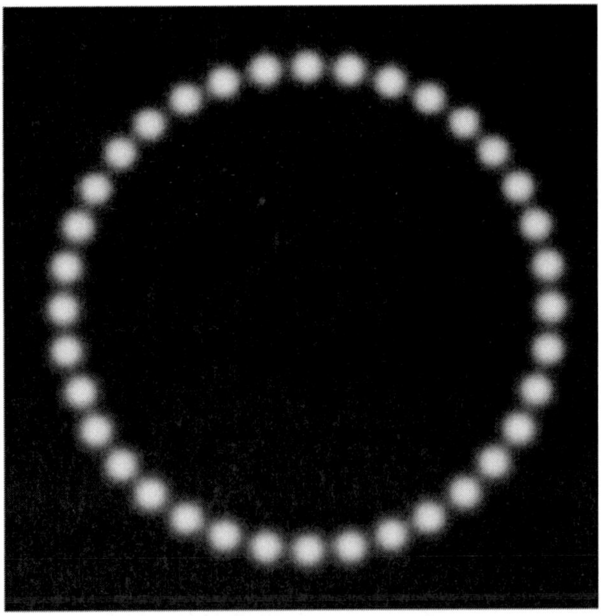

Figure 6. Modes of Strömgren's shell.

CREIL effect along the path in gas to the surface.

- Approaching the surface, the emission grows exponentially, while the redshift decreases (Figure 7(a)).

- Above the surface, the light passes through a region in which the irradiance of superradiant UV is large, so that the frequencies of the low radiance spectrum are increased (Figure 7(b)). In terms of increasing wavelength, the emission line appears suddenly at a wavelength $\lambda_1 < \lambda_0$, then decreases exponentially. In practice, the observation adds distinct spectra of rays, so that the intensity growth is less abrupt.

3.3. Absorption of a Continuous UV-X Spectrum by Relatively Cold (30000 K) Atomic Hydrogen

Suppose that a star is surrounded by hydrogen whose temperature and pressure decrease with altitude. Follow the light emitted by the star.

The star emits a high frequency continuous spectrum. Close to the star, hot, excited atomic hydrogen emits the Lyman lines which are much broadened by the pressure.

The excited states are nearly depopulated by a decrease of temperature under 50,000 K. The absorption at Lyman frequencies prints the spectrum in light. The pressure drop sharpens the absorbed spectral lines. The absorption of Lyman α line populates the 2P states of the atoms, so that, where the duration of the free path of the atoms becomes larger than a nanosecond, a CREIL appears, which redshifts continuously the frequencies of light. The fast passing of the absorbing lines through the spectrum absorbs little light, so that the broad, weakly absorbed lines are invisible.

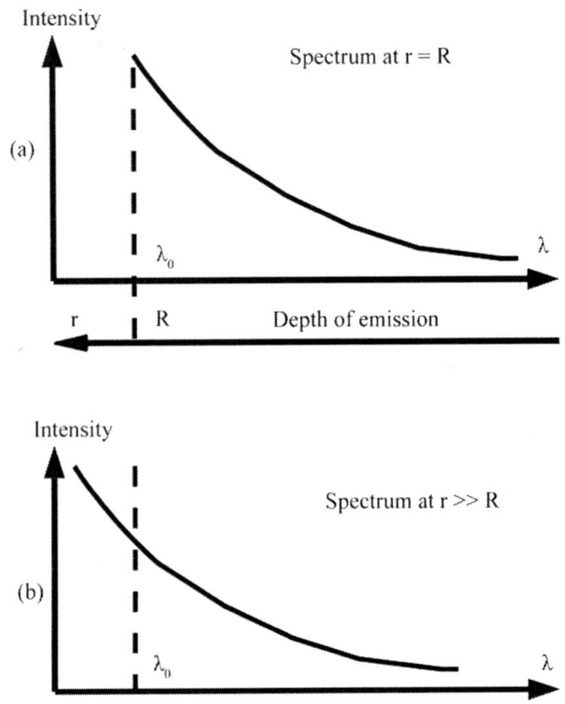

Figure 7. Spontaneous emission of Lyα line on a ray observed inside the main ring of SNR1987A.

However, if an absorbed line reaches by redshift the Lyman α frequency, 2P atoms are not created, the redshift stops, so that the whole absorption spectrum of H is written clearly in the light.

But the stop is not perfect because excited $n>2$ states provide a weak redshift and may decay into $n=2$ states: the redshift restarts.

Thus, two Lyman spectra are written in the light, one at its regular frequencies, the other with the redshift $Z_{(\beta,\alpha)}=\left(\nu_\beta-\nu_\alpha\right)/\nu_\alpha$ which puts the Lyman β line to the α. The process may restart, in which the lines of both absorbed spectra may generate the absorption of a new spectrum.

The final result is a set of absorption lines in which all lines result from the absorption of Lyman spectra redshifted by successive redshifts which shift various Lyman lines to Lyman α . Neglecting the dispersion of the redshifts, the absorption frequencies of the lines are the Lyman frequencies shifted by $b\times Z_{\beta,\alpha}\times c\times Z_{\gamma,\alpha}\times\cdots$ where $b,c,\cdots$ are positive integers This formula may be simplified by a computation of approximate values of $Z_{(\beta,\alpha)}$ and $Z_{(\gamma,\alpha)}$:

$$\begin{aligned}Z_{(\beta,\alpha)} &= \left(\nu_\beta-\nu_\alpha\right)/\nu_\alpha\\ &= \left[\left(1-1/3^2\right)-\left(1-1/2^2\right)\right]/\left(1-1/2^2\right)\\ &\approx 5/27\approx 0.1852\approx 3\times 0,0617\end{aligned}$$

$$\begin{aligned}Z_{(\gamma,\alpha)} &= \left(\nu_\gamma-\nu_\alpha\right)/\nu_\alpha\\ &= \left[\left(1-1/4^2\right)-\left(1-1/2^2\right)\right]/\left(1-1/2^2\right)\\ &= 1/4 = 0.25 = 4\times 0,0625;\end{aligned}$$

In the last formula, taking into account that the gamma line is weaker than the beta, we can approximate 0.0625 by 0.0617. Thus, the shifts are approximately $(3b+4c)\times 0.0617$. 0.0617 is very close to the experimental value of the Karlsson's constant 0.061.

Some values of $n=3b+4c$ correspond to superposed, therefore intense lines: n = 10 = 3b + 4c = 3 + 3 + 4 = 3 + 4 + 3 = 4 + 3 + 3.

3.4. Frequency Shifts of Spectra Emitted inside Stars

Compression due to the huge surface gravity of a star similar to the Sun, leads at a short distance below the surface, despite the high temperature, the hydrogen to a quasi-crystalline state. Fixed atoms radiate sharp lines, so CREIL is possible.

Relatively heavy atoms can be at a temperature high enough to emit X-UV rays which are redshifted while they propagate to the surface. Light emitted by lighter atoms at a lower depth and frequency, may exchange more energy with the radiation of heavier atoms than with thermal radiation, so that their frequency is increased. Emissions near the surface are not affected by CREIL.

Frequency shifts are more intense at the limb than at the center of the disk because the path in "crystallized" hydrogen atoms is larger.

4. PLAUSIBLE ASTROPHYSICS

4.1. Brilliant Circles

The Hubble picture HEIC 1018A (Figure 8) shows an almost perfect circle that can be the image of the limb of a Strömgren sphere.

4.2. Rings of Supernova 1987A

If the star which creates a Strömgren sphere has satellites or a ring in its equatorial plane, the radiation is partially absorbed, so the cooling of the plasma is increased in this plane, the sphere is choked and takes the form an hourglass.

When the star whose name became SN1987A was lit, the astronomers observed light echoes reflected by clouds of hydrogen. The variation of the brightness of the star was used to compare the propagation time of the direct light and reflected light, so that the geometry of these clouds has been drawn in the shape of an hourglass.

Then came three rings (Image heic0704a from Hubble, Figure 9) at the same time as the star disappeared.

Martin and Arnett [11] have drawn a rough cut of the hourglass (Figure 10) to show that lines tangent to the hourglass, directed towards the Earth, draw diameters of the rings. These authors were criticized by

Burrows et al. [12] who, disregarding the superradiance, got an image like Figure 4, rather than Figure 6. Coherent spectroscopy explains the ring structure, on the other hand their brightness at the expense of that of the star.

Figure 8. Heic 1018a (Hubble).

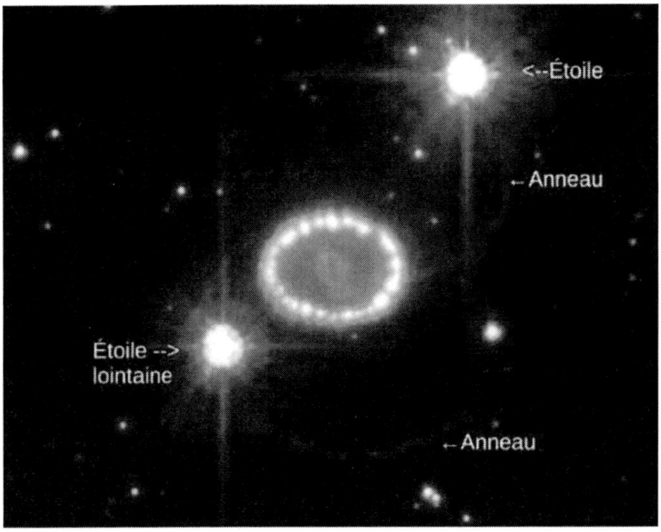

Figure 9. SNR1987A heic 0704a (ESA/Hubble).

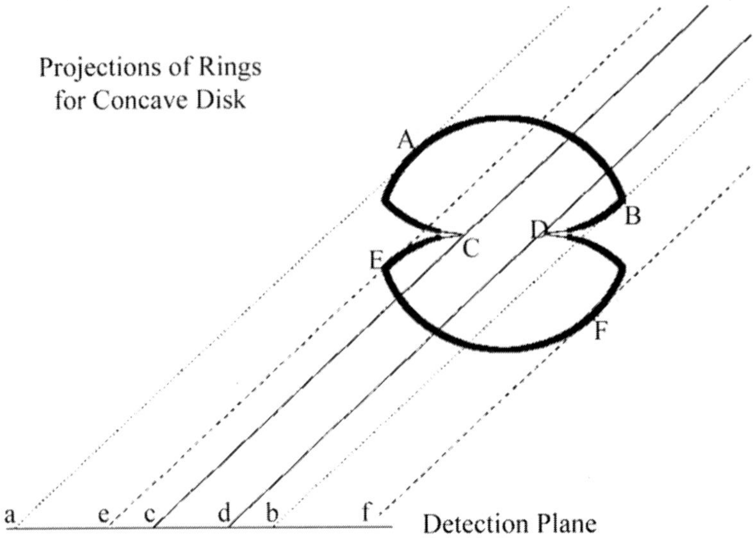

Figure 10. The rings of SNR1987A are the limbs of the hourglass. From Martin & Arnett [11].

Many systems show sets of points interpreted by gravitational lensing, e.g. Einstein Cross. This interpretation has two problems:

- The number of alignments of heavy objects seems too large;
- The even parity of the number of points (center point excluded) in clean systems is not explained.

4.3. Lineshapes of Spectra Emitted inside an Hydrogen Plasma, Seen, for Instance, inside the Smallest Ring of SNR1987A

The spectrometer adds intensities of rays emerging from the sphere under various incidences. These rays have slightly different spectra, so that the experimental maximum (Figure 11) is less sharp than on Figure 7. This typical lineshape is often observed for lines emitted by various atoms in an hydrogen plasma.

The maximum intensity in Figure 7 is limited because the exponential is cut by the exit of the ray from the sphere. Instead, the calculation of Monte-Carlo (Figure 12) shows a maximum intensity limited by the saturation of the curve tracer.

4.4. Spectroscopic Explanation of Hubble's Law

Assume that the density of excited atomic hydrogen H* and that the electromagnetic background irradiances are constant in the low pressure regions of universe. Neglecting the dispersion of CREIL, the redshift of the light is proportional to the column density of H*: Hubble's law measures with a good approximation the column density of n = 2 atomic hydrogen. In cold regions of the nebulae, the density of H* seems constant, so that Hubble's law measures a distance. But the density of H* may be large close to hot objects.

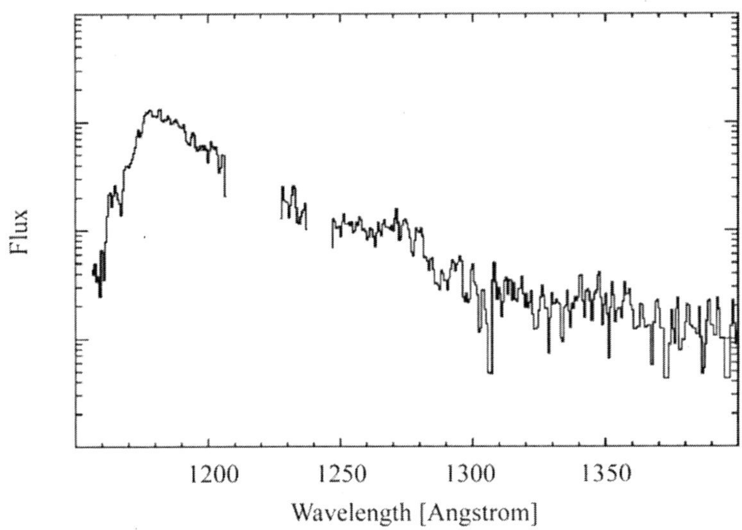

Figure 11. Experimental spectrum inside the small ring of SNR1987A by Michael et al. [13].

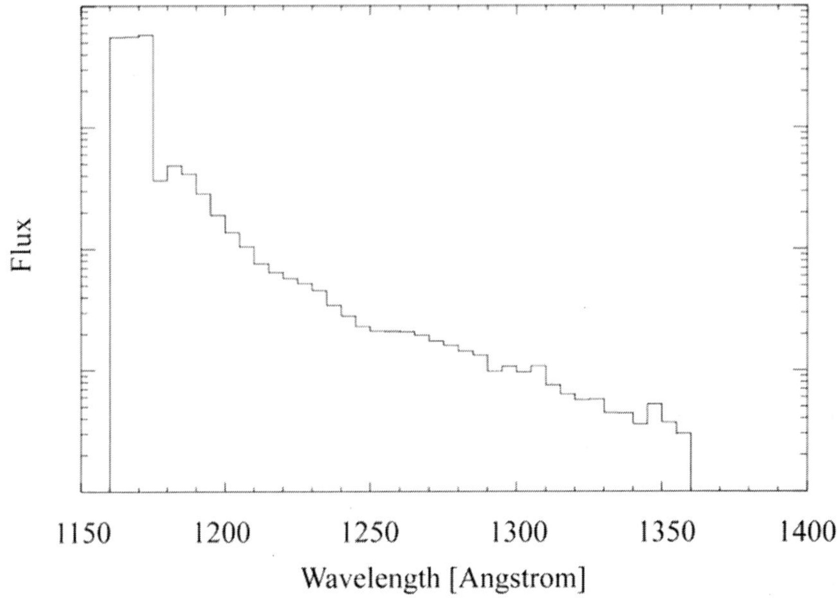

Figure 12. Spectrum computed by Michael et al. using a Monte-Carlo program [13].

Many astronomers, Halton Arp in particular [14], remarked that sets of stars which seem bound may have different redshifts, the hottest being the most redshifted. As hydrogen plasma often contain a large density of H*, some people try to explain their power to redshift light by recoil effects on free protons or electrons, but there is no hope to find such an effect in an homogeneous plasma.

4.5. Intergalactic Voids

Some galaxies contain massive and hot stars emitting hot hydrogen winds. Some winds remain excited about until some other galaxies. The bubbles of H* thus formed are swollen on maps by a blind use of Hubble's law. The radial distances of galaxies maps and transverse distances that are deducted are exaggerated. This results in a miscalculation of the dynamics of spiral galaxies, corrected by the improbable hypothesis of dark matter.

From a local perspective, the bubbles increase irregularly radial distances, generating voids which contain few stars, creating an observed spongy structure in the maps of galaxies.

4.6. Experimental Observation of CREIL

This check requires a long light path in H* at low pressure. A rough calculation of adiabatic expansion of hydrogen plasma $\left(p^{+}+e^{-}\right)$ making the solar wind produces H* at 10 AU, then de-excites it at 15 AU. It is a weak type of Strömgren shell.

Sunlight is redshifted as it passes through this shell, producing a pulsed excitation of the spin-coupling states of H* atoms, excitation detected by a continuous wave probe beam.

The carrier of the signals exchanged with the Pioneer 10 and 11 probes undergoes an increase in frequency during its passage through the layer containing H* (Figure 13). Note that this frequency increases during the pulses of the sunlight, with the destruction of the temporal coherence while the frequency changes. The detector extracts a continuous wave from the noisy one it receives. Its frequency is increased.

This interpretation could be tested by not measuring the frequency of a carrier, but the frequency of pulses which would be affected by a Doppler effect, but not by a CREIL. The experiment may be very long because the frequency shift is almost constant beyond 15AU. Thermal radiation observed at 2.7 K can be amplified by the same mechanism. Its anisotropy observed related to the ecliptic, is explained by the anisotropy of the solar wind.

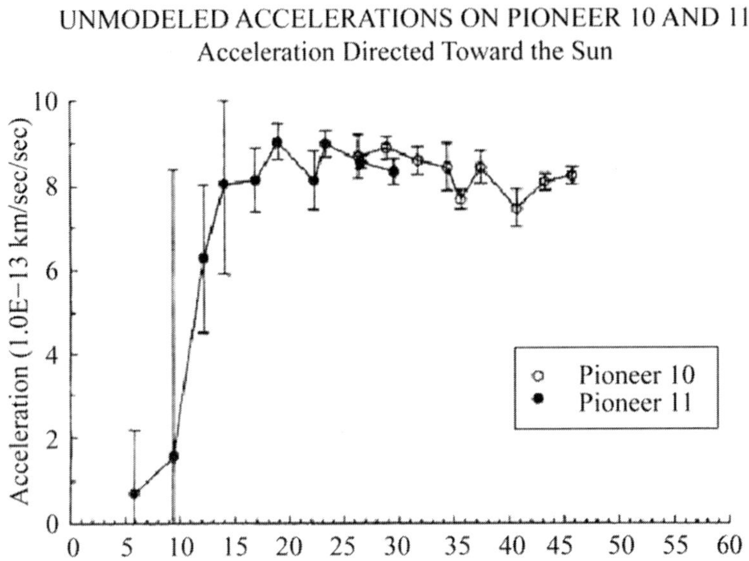

Figure 13. Pioneer 10 and 11 anomalous acceleration from Anderson et al. [15].

4.7. Periodicities of the Redshifts of Quasars and Galaxies

The theory in 3.3 explains the periodicities observed in the spectra of the quasars: the propagation of light is divided into cycles which during the one hand redshift light, and, on the other hand, whenever an absorbed line gets the Lyman α frequency, inscribe emission and absorption lines of the gas. These cycles stop where the local gas is not anymore able to produce low pressure 2P hydrogen, at any time of a cycle. Considering a large set of observed objects, the probability to observe an object whose final redshift corresponds to a stop by coincidence of an absorbed line with the Ly α line is large.

In 1968, Burbidge [16] then many other authors found experimentally this result: The redshifts of the galaxies are often given by $Z_{obs} \approx 0.061n$ where n has certain integer values, larger than 2, and 0.061 is the Karlsson periodicity.

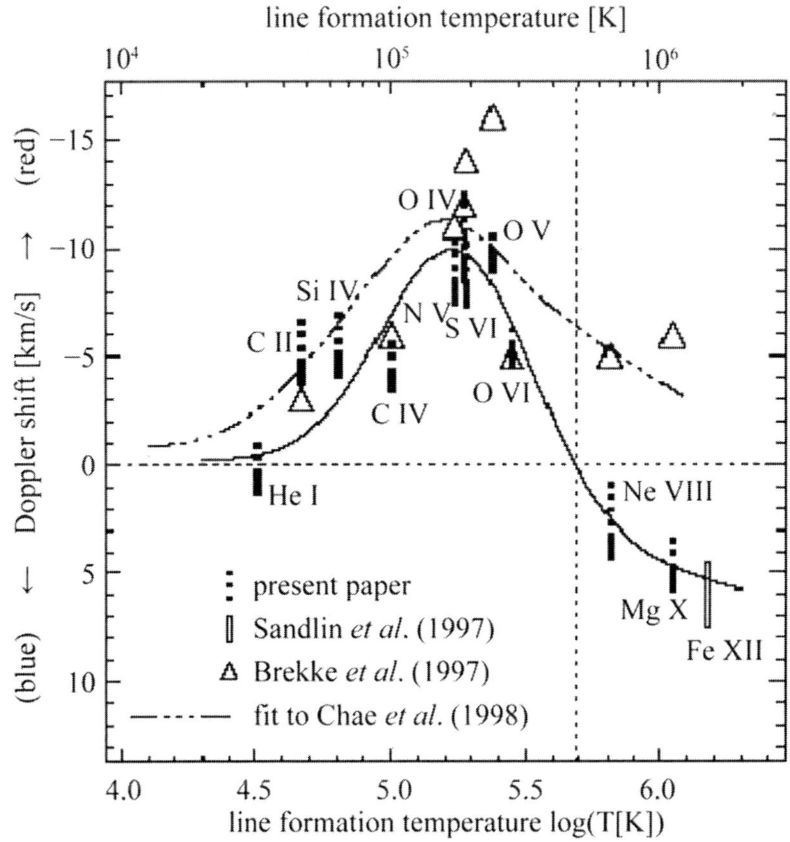

Figure 14. UV-X spectrum of the Sun [15].

4.8. SOHO Observation of UV-X Spectrum of the Sun

Peter & Judge [17] attribute the frequency shifts of XUV rays of Sun to Doppler effect produced by a vertical velocity of the emitting gas. With this hypothesis, the maximum effect is at the center of the disc, and it is null at the limb whose radiation is used to recalibrate the spectrometer. But, using the internal calibration, the frequencies observed at the limb are further from laboratory frequencies than those observed at the center.

Figure 14 reproduces a figure of paper [17]. With the theory given in 3.4, the meaning of the y-axis of this figure and its scale must be changed because this figure represents the difference between the frequency shift (minimum) at the center of the disc and the frequency shift (maximum) at the limb. Thus, the internal calibration appears good.

5. CONCLUSIONS

Most physicists have long held that the coherent interactions of light and matter are negligible, with the exception of refraction. Little affected by the discovery of the laser, astrophysicists have kept this vision of coherent interactions expressed by Menzel in 1931 and confirmed by an aberrant concept of the photon as a small particle able to interact only with a single atom in a gas at low pressure. This conception, incompatible with quantum electrodynamics, has been strongly criticized by W. E. Lamb Jr., W. P. Schleich, M. O. Scully and C. H. Townes [1].

Frequency variations are observed in short pulses from femtosecond lasers. It is "Impulsive Stimulated Raman Scattering" (ISRS). These variations are approximately inversely proportional to the cube of the pulse duration. Replacement of femtosecond pulses by pulses 10^5 times longer that model natural incoherent light, reduces the frequency shift by a factor 10^{15}, so that an observation requires an astronomical path.

This ISRS in incoherent natural light requires a gas having a quadrupole resonance frequency below 1 GHz, but as high as possible. Atomic hydrogen in excited levels 2S or 2P (denoted H*) is suitable. It is found in hydrogen plasma, and 2P H* is formed in cold atomic hydrogen by Lyman α optical pumping. In practice, several effects must be coupled into a parametric effect named Coherent Raman Effect on Incoherent Light (CREIL) so that the Raman levels are not excited. The CREIL exchanges progressively energy between beams of light, in accordance with thermodynamics.

The use of optical coherence allows to build models that appear to explain many observations, in particular:

- A bright star with a ring or satellites and immersed in a cloud of hydrogen has the appearance, the hydrogen spectrum and the evolution of SNR1987A;
- It is surprising that not a single "accreting neutron star" has been seen in close hydrogen clouds, when they should be bright. In fact many have been observed, but they are called "quasars";
- The anomalous acceleration of Pioneer 10 and 11 probes is a consequence of the blueshift of the microwave carrier of the signals exchanged between the probe and Earth: The adiabatic expansion of the solar wind creates H* between 10 and 15 AU, which catalyzes a transfer of energy from incoherent sunlight to microwaves;
- The strong reddening of light in the vicinity of hot stars that produce a large amount of H*, are usually interpreted as voids, so that the maps of galaxies are spongy;
- A quantitative result is the precise, theoretical computation of Karlsson's constant, measured from statistics of the redshifts of stars and quasar spectra. Other theories are unable to explain the origin of this constant, thus to compute its value.

Our models must be improved, introducing gas other than hydrogen, making them more quantitative; other models may be easily set to study other observations: There remains a lot of work to develop our start of interpretation of astronomical observations using optical coherence.

Optical coherence could be an alternative to the Big Bang theory. But many years seem necessary for the adoption of the upheaval that is the substitution of a basic, experimentally verified theory, to a wonderful theory.

REFERENCES

1. W. E. Lamb, Jr., W. P. Schleich, M. O. Scully and C. H. Townes, Reviews of Modern Physics, Vol. 71, 1999, pp. S263-S273.

2. D. H. Menzel, PASP, Vol. 43, 1931, pp. 70-74.

3. Z. Zheng, The Astrophysical Journal, Vol. 578, No. 1, 2002, pp. 33-43.

4. A. Einstein, Physikalische Zeitschrift, Vol. 18, 1917, pp. 121-128.

5. M. Planck, Deutsche Physikalische Gesellschaft, Vol. 13, 1911, pp. 138-175.

6. Y.-X. Yan, E. B. Gamble Jr. and K. A. Nelson, Journal of Chemical Physics, Vol. 83, No. 11, 1985, pp. 5391-5399.

7. L. Dhar, J. A. Rogers and K. A. Nelson, Chemical Reviews, Vol. 94, 1994, pp. 157-193.

8. I. H. Lee, K. J. Yee, K. G. Lee, E. Oh, D. S. Kim and Y. S. Lim, Journal of Applied Physics, Vol. 93, 2003, pp. 4939-4941.

9. G. L. Lamb Jr., Reviews of Modern Physics, Vol. 43, 1971, pp. 99-124.

10. B. Strömgren, The Astrophysical Journal, Vol. 89, 1939, pp. 526-547.

11. C. L. Martin and D. Arnett, The Astrophysical Journal, Vol. 447, 1995, pp. 378-390.

12. C. J. Burrows, J. Krist, J. J. Hester, R. Sahai, J. T. Trauger, K. R. Stapelfeldt, J. S. Gallagher III, G. R. Ballester, S. Casertano, J. T. Clarke, D. Crisp, R. W. Evans, R. E. Griffiths, J. G. Hoessel, J. A. Holtzman, J. R. Mould, P. A. Scowen, A. M. Watson and J. A. Westphal, The Astrophysical Journal, Vol. 452, 1995, pp. 680-684.

13. E. Michael, R. McCray, R. Chevalier, A. V. Filippenko, P. Lundqvist, P. Challis, B. Sugerman, S. Lawrence, C. S. J. Pun, P. Garnavich, R. Kirchner, A. Crotts, C. Fransson, W. Li, N. Panagia, M. Phillips, B. Schmidt, G. Sonneborn, N. Suntzeff, L. Wang and J. C. Wheeler, The Astrophysical Journal, Vol. 593, 2003, pp. 809-830.

14. H. Arp, "Quasars Redshifts and Controversies," Cambridge University Press, Cambridge, 1989.

15. J. D. Anderson, P. A. Laing, E. L. Lau, A. S. Liu, M. M. Nieto and S. G. Turyshevet, "Study of the Anomalous Acceleration of Pioneer 10 and 11," 2005, arXiv:gr-qc/ 0104064.

16. G. Burbidge, The Astrophysical Journal, Vol. 154, 1968, pp. L41-L48.

17. H. Peter and P. G. Judge, The Astrophysical Journal, Vol. 522, 1999, pp. 1148-1166.

Chatper 4

Characterization of MHz pulse repetition rate femtosecond laser-irradiated gold-coated silicon surfaces

Manickam Sivakumar[1,3*], Krishnan Venkatakrishnan[2], Bo Tan[1]

(1) Department of Aerospace Engineering, Ryerson University
(2) Department of Mechanical and Industrial Engineering, Ryerson University
(3) On leave from Department of Sciences, Amrita School of Engineering, Amrita Vishwa Vidyapeetham

ABSTRACT

In this study, MHz pulse repetition rate femtosecond laser-irradiated gold-coated silicon surfaces under ambient condition were characterized by scanning electron microscopy (SEM), transmission electron microscopy (TEM), X-ray diffraction analysis (XRD), and X-ray photoelectron spectroscopy (XPS). The radiation fluence used was 0.5 J/cm^2 at a pulse repetition rate of 25 MHz with 1 ms interaction time. SEM analysis of the irradiated surfaces showed self-assembled intermingled weblike nanofibrous structure in and around the laser-irradiated spots. Further TEM investigation on this nanostructure revealed that the nanofibrous structure is formed due to aggregation of Au-Si/Si nanoparticles. The XRD peaks at 32.2°, 39.7°, and 62.5° were identified as (200), (211), and (321) reflections, respectively, corresponding to gold silicide. In addition, the observed chemical shift of Au *4f* and Si *2p* lines in XPS spectrum of the irradiated surface illustrated the presence of gold silicide at the irradiated surface. The generation of Si/Au-Si alloy fibrous nanoparticles aggregate is explained by the nucleation and subsequent condensation of vapor in the plasma plume during irradiation and expulsion of molten material due to high plasma pressure.

INTRODUCTION

Nanostructures of Au, Au-Si alloy, and Si have been employed in micro and nanoelectromechanical systems [1], biosensors [2], and photonics [3, 4]. The field-emission property [5] of Au-Si alloy structures is used for the fabrication of panels and displays. Au-Si alloy nanoparticles are increasingly relevant as they are used as catalysts in the growth of nanowires [6]. Recently, femtosecond lasers have proven to be a powerful tool for nanostructuring of bulk metals [7–10]. Femtosecond laser pulses have also been used for precise nanostructuring of thin films with minimal thermal side effects [11–13]. The ultrafast excitation of materials controls the deposited energy in the material with femtosecond pulses. As a result, nanostructures with spatial resolution smaller than the wavelength of radiation can be generated. Although interaction of laser radiation with gold, gold-silicon thin films that lead to the formation of microbumps/nanojet structures have been studied [11, 14, 15], investigations on the Au-Si/Si fibrous nanoparticles aggregate formation using femtosecond laser radiation under ambient condition have not been reported. In the previous studies, synthesis of self-assembled weblike fibrous nanoparticles aggregate structures, nanofibers, and nanoscale tips with bulk semiconductor, metallic, and dielectric materials using femtosecond laser radiation under ambient condition was reported [8, 16–18]. The fibrous structure generation is explained by nucleation and condensation of plasma plume generated during the irradiation process. It was comprehended that generation of nanofibrous structures can significantly be controlled by laser radiation fluence, laser interaction time, and pulse repetition rate [16]. This study is aimed to investigate the synthesis of Si/Au-Si alloy nanoparticles aggregate with femtosecond laser irradiation of Au-coated silicon samples and the influence of laser interaction time on composition of structures. The plausible mechanism underlying Au-Si alloy nanoparticles aggregate generation will also be discussed. The irradiated sample surfaces are characterized using scanning electron microscopy (SEM), transmission electron microscopy (TEM), and X-ray diffraction analysis (XRD). The chemical composition of nanoparticles aggregate is further analyzed using X-ray photoelectron spectroscopy (XPS).

EXPERIMENTAL METHODS

The laser source used (λ = 1030 nm) is a direct-diode pumped Yb-doped fiber oscillator/amplifier system capable of delivering a maximum of 15 W average power with a pulse repetition rate ranging from 200 kHz to 25 MHz. The beam profile is Gaussian and the spot diameter is (10 μm) measured at $1/e^2$. The samples used were gold-coated (thickness 200 to 400 nm) silicon wafers. The laser beam is focused on the sample surface with a lens of focal length of 71 mm and scanned using a computer-controlled galvanometer to produce arrays of spots. The experiments were carried out in air at atmospheric pressure. The radiation fluence used was 0.5 J/cm^2 at a pulse repetition rate of 25 MHz with an interaction time of 1 ms and with the pulse width of 214 fs. Three sets of samples were prepared with same experimental conditions. One set was used for SEM analysis. The other set was used for XRD analysis followed by TEM investigations. Third set was used for XPS analysis. XRD measurements were performed with a Cu Kα radiation (λ = 0.154184 nm). The diffractograms were recorded using Bruker detector from 20° to 70°. To transfer the nanostructures to TEM grids, the samples were sonicated in isopropanol solution. A drop of the nanoparticles aggregate dispersed solution was placed on the copper grid and allowed to dry in air. XPS measurements were carried out on a Thermo Scientific K-Alpha

XPS spectrometer. A monochromatic Al Kα X-ray source was used, with a spot area (on a 90° sample) of 400 μm.

RESULTS AND DISCUSSION

SEM micrographs of the sample surfaces around the laser-irradiated spots are shown in Figure 1a. Weblike fibrous nanoparticles aggregate with certain degree of porosity is observed in and around the laser-irradiated spots with all processing parameters used. SEM/EDX analysis of the nanostructure shows the presence of gold, silicon, and oxygen. In addition to Au-Si, nanoparticles of amorphous silicon which are comparatively smaller are also observed (Figure 1b). The size of Au-Si nanoparticles in the fibrous nanoparticles aggregate structure is bigger than that of Si particles as evidenced from TEM analysis (Figure 1b, c). Moreover, the different nanoparticles are intermixed with each other in the aggregate. TEM/EDX analysis of the nanoparticles aggregate revealed the composition Au-Si nanoparticles. The relative amount of silicon and gold in Au-Si nanoparticle varies with different particles (Figure 2).

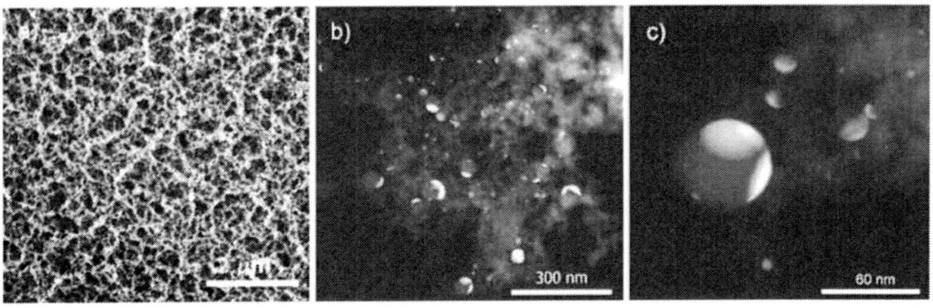

Figure 1 SEM and TEM micrographs of the sample surface irradiated at laser radiation fluence (0.5 J/cm ^{2}) with interaction time (1 ms) and the nanoparticles aggregate respectively. **(a)** surface featuring weblike nanoparticles around the laser spot, **(b, c)** Si/Au-Si nanoparticles in the aggregate structure.

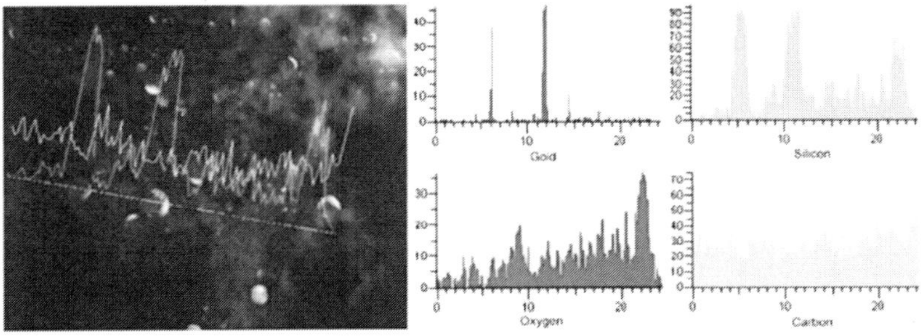

Figure 2. TEM-EDX analysis of the nanoparticles in the aggregate structure.

X-ray diffractograms of both treated and untreated samples were performed with Cu Kα radiation (λ = 0.1541848 nm) [19] (Figure 3). The peaks 32.2°, 39.7°, and 62.5° were identified as (200), (211), and (321) reflections, respectively, corresponding to JCPD file for Au81Si19 (JCPD 39-0735). Further peaks at 38.2°, 44.38°, and 64.6° were identified to originate from (111), (200), and (220) planes of Au (JCPD 04-0784), respectively. The average Au-Si particles size is calculated from the full-width at half-maximum (FWHM) of the diffraction peaks using the Debye-Scherrer formula [20] $D = k\lambda/\beta\cos\theta$, where D is the mean grain size, k is a geometric factor (= 0.89), λ is the X-ray wavelength, β is the FWHM of diffraction peak, and θ is the diffraction angle. The grain sizes of Au-Si calculated from the peaks 32.2° and 39.7° are 26 and 55 nm, respectively.

The chemical state of Au, Si, and oxygen atoms for both untreated and laser-treated samples was investigated by XPS. The correction of the XPS spectra for the charge accumulation was performed using C 1s peak (BE = 284.6 eV), which can be ascribed to contaminant hydrocarbons. Figure 4 shows XPS spectra of Au 4f line obtained for the untreated and laser-treated samples. For the untreated samples, the peaks of Au $4f_{7/2}$ and Au $4f_{5/2}$ lines are located at BE = 83.82 and 87.4 eV, which correspond to elemental gold. Laser-irradiated samples showed broadened peaks of Au $4f_{7/2}$ and Au $4f_{5/2}$ lines. Deconvolution of these lines showed the presence of two peaks in each line. For instance, the peaks of Au 4*f* $_{7/2}$ line appear at 85.33 eV, which is a characteristic of metallic gold, and at 83.9 eV, 1.43 eV higher in BE can be ascribed to silicide [21–23]. Moreover, the 83.9 eV peak is due to the interaction of gold with silicon at the interface [24] by laser irradiation, resulting the formation of gold silicide in the nanoparticles aggregate. The presence of elemental gold with laser-irradiated samples is due to untreated areas of the sample around the laser spot above which nanoparticles aggregate was formed. Since the thickness of gold layer deposited on sample surface is about 200 to 400 nm, XPS analysis of untreated sample has not showed the silicon peak. However, investigations on the laser-irradiated sample surface revealed the presence of Si 2*p* line. The deconvoluted spectrum of the laser-treated sample is shown in Figure 5. The peaks are due to silicon nanoparticles in the aggregate structure.

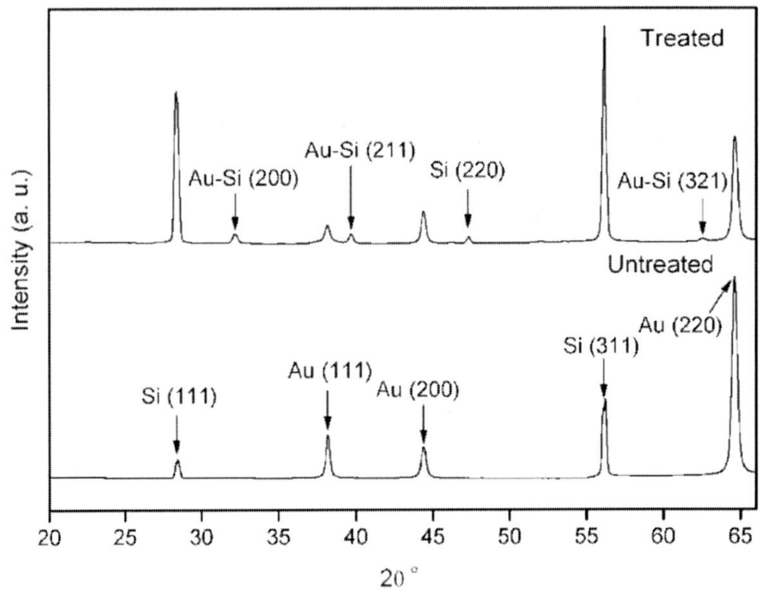

Figure 3. X-ray diffractograms of both treated and untreated samples. The peak at 39.7° in the treated sample is attributed to the Au-Si alloy phase.

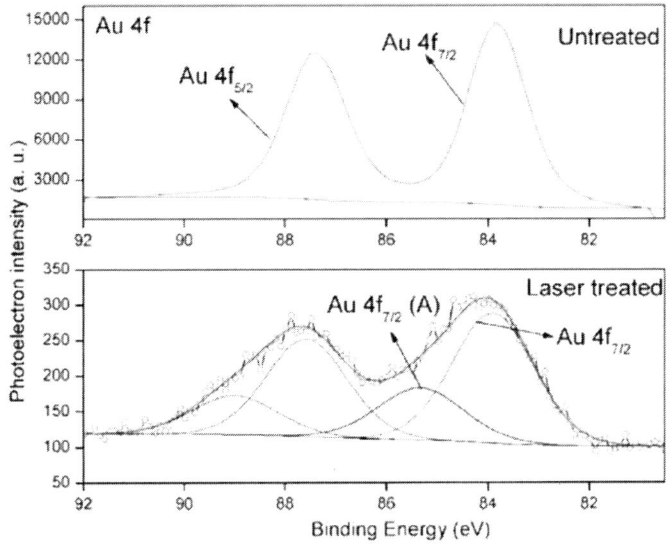

Figure 4. XPS spectra of untreated and laser treated samples for Au 4 **f** system**.**

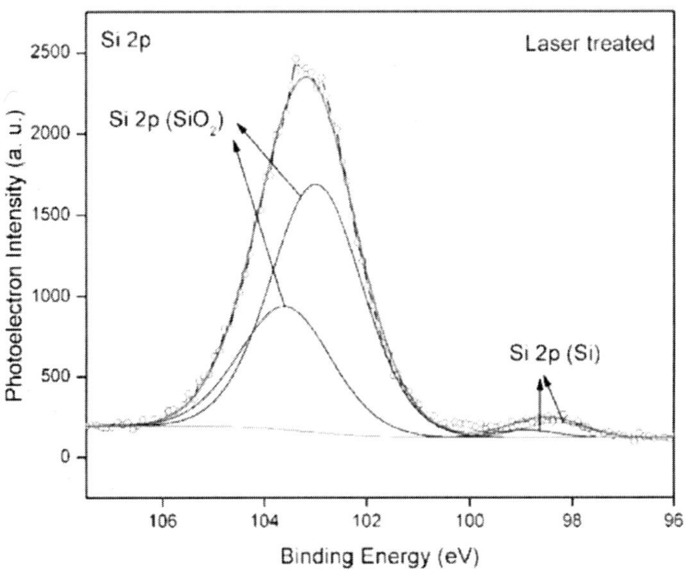

Figure 5. XPS spectra of laser treated samples for Si 2 **p** system**.**

Irradiation of metal films using femtosecond laser radiation results in fast nonequilibrium processes such as laser melting and film disintegration [7, 25]. The energy from laser radiation is absorbed by the conduction band electrons and results in a sharp increase in electronic temperature near the irradiated front surface. Since the heat capacity of electrons in a metal is much smaller than that of lattice, an ultrashort laser pulse can heat electrons to a very high temperature while leaving the lattice relatively cool. The fast temperature-dependent electron heat conduction leads to the redistribution of the deposited energy within the film. This process occurs simultaneously with a more gradual energy transfer from electrons to the lattice vibrations due to electron-phonon coupling. The time duration of energy transfer and the ensuing equilibrium processes depends on electron-phonon coupling in gold. For Au, the coupling is 2.1×10^{16} W/m^3 K, and the energy is transferred to the lattice within 15 ps. The equilibrium between hot electrons and lattice takes place with a time limit of up to 50 ps [25]. During irradiation, first few pulses alter the gold film and significantly increasing the absorption of this modified surface for the subsequent pulses [9]. At 25 MHz pulse repetition rate, the pulse separation time is 40 ns, Au film reaches high temperature due to accumulation of heat from successive pulses. Laser pulse repetition rate plays a significant role in generating these structures due to cumulative heating. At MHz laser pulse repetition rate, the delay between successive pulses is comparable to the critical time of nucleation [8]. Besides, repetition rate helps to sustain the molten liquid thereby maintaining the plasma and nanoparticles agglomeration. Taking into consideration the different plume components, formation of gold nanoparticles with various sizes under femtosecond laser irradiation is explained by nucleation and condensation of vapor in the plasma plume and explosion of molten material due to high plasma pressure [17].

The melting point of Au is 1063°C while for Si it is 1414°C. Although for an alloy which contains 81% Au and 19% Si, the melting point is 359°C, called the eutectic point. The formation of Au-Si alloy catalyst is explained via a Vapor-Liquid-Solid (VLS) process [6] during the synthesis of Si nanowires. In this experiment, since the laser radiation fluence (0.5 J/cm^2) used is much above the ablation threshold of Au (0.2 J/cm^2) with multiple pulses [26], gold may diffuse into the silicon substrate to form an alloy at the interface [27]. This alloy layer at the interface started melting due to cumulative heating by the subsequent laser pulses. The expulsion of molten alloy material results in the formation of alloy nanoparticles. Once the alloy layer is depleted further irradiation ablates the underlying silicon substrate and generates the plasma. At this point, the silicon nanoparticles are generated by nucleation and condensation of vapor in the plasma [8] and agglomerates with Au-Si alloy nanoparticles to form weblike nanofibrous structure. Although irradiation of molten alloy nanoparticles by subsequent laser pulses may increase their temperature, it is not supporting the growth of silicon nanowires [27]. The relative proportion of alloy nanoparticles in the aggregate nanostructure is mainly determined by the laser interaction time. This is implicitly understood from the SEM/EDX analysis, which shows the atomic percent of gold and silicon as a function of interaction time in the fibrous aggregate structure. In contrast to normal VLS process [6], where the source of semiconductor is supplied as vapor phase, in this case both gold and silicon are originated from the gold-coated solid silicon substrate. Upon laser irradiation, molten Au-Si alloy layer is formed and transforms into molten nanoparticles and agglomerates as solid weblike self-assembled fibrous structure. In other words, the fibrous nanoparticles aggregate process can be regarded as Solid-Liquid-Solid [27] process. In this experiment, Si/Au-Si fibrous nanoparticles aggregate is generated in a single step under ambient condition. The existence of Au-Si alloy in the nanoparticles aggregate is corroborated from XRD, TEM/EDX, and XPS analysis. Moreover, the size of alloy nanoparticles calculated from the peaks of XRD diffractogram matches well with the size observed from TEM micrographs.

CONCLUSIONS

A simple method of generating Si/Au-Si alloy nanoparticles aggregate using MHz pulse repetition rate femtosecond laser radiation under ambient conditions is reported. The generation of nanoparticles aggregate is explained by nucleation and condensation of vapor in the plasma plume and expulsion of molten material due to high plasma pressure. This technique could be extended to generate other metal-semiconductor alloy nanostructures. Further studies are required to find the influence of film thickness and laser processing parameters on nanostructure generation.

ABBREVIATIONS

SEM: scanning electron microscopy

TEM: transmission electron microscopy

VLS: vapor-liquid-solid

XPS: X-ray photoelectron spectroscopy

XRD: X-ray diffraction analysis.

ACKNOWLEDGEMENTS

This study was funded by the Natural Science and Engineering Research Council of Canada.

AUTHORS' CONTRIBUTIONS

SM carried out laser processing of the samples, characterisation and drafted the manuscript. KV conceived of the study, and participated in its design and co ordination. BT conceived of the study, and participated in its design and co ordination. All authors read and approved the final manuscript.

REFERENCES

1. Cheng YT, Lin LW, Najafi K: Localized silicon fusion and eutectic bonding for MEMS fabrication and packaging. J Microelectromech Syst 2000, 9: 3–8.

2. Lesuffleur A, Im H, Lindquist NC, Oh SH: Periodic nanohole arrays with shape-enhanced plasmon resonance as real-time biosensors. Appl Phys Lett 2007, 90: 243110.

3. Mahmood AS, Sivakumar M, Venkatakrishnan K, Tan B: Enhancement in optical absorption of silicon fibrous nanostructure produced using femtosecond laser ablation. Appl Phys Lett 2009, 95: 034107.

4. Bettotti P, Cazzanelli M, Dal Negro L, Danese B, Gaburro Z, Oton CJ, Prakash GV, Pavesi L: Silicon nanostructures for photonics. J Phys Condens Matter 2002, 14: 8253–8281.

5. Wan Q, Wang TH, Lin CL: Self-assembled Au-Si alloy nanocones: synthesis and electron field emission characteristics. Appl Surf Sci 2004, 221: 38–42.

6. Lu W, Lieber CM: Semiconductor nanowires. J Phys 2006, 39: R387-R406.

7. Hwang TY, Vorobyev AY, Guo CL: Ultrafast dynamics of femtosecond laser-induced nanostructure formation on metals. Appl Phys Lett 2009, 95: 123111.

8. Tan B, Venkatakrishnan K: Synthesis of fibrous nanoparticle aggregates by femtosecond laser ablation in air. Opt Express 2009, 17: 1064–1069.

9. Vorobyev AY, Guo C: Enhanced absorptance of gold following multipulse femtosecond laser ablation. Phys Rev B 2005, 72: 195422.

10. Vorobyev AY, Guo CL: Femtosecond laser nanostructuring of metals. Opt Express 2006, 14: 2164–2169.

11. Kuznetsov AI, Koch J, Chichkov BN: Nanostructuring of thin gold films by femtosecond lasers. Appl Phys A 2009, 94: 221–230.

12. Miyaji G, Miyazaki K: Origin of periodicity in nanostructuring on thin film surfaces ablated with femtosecond laser pulses. Opt Express 2008, 16: 16265–16271.

13. Ivanov DS, Rethfeld B, O'Connor GM, Glynn TJ, Volkov AN, Zhigilei LV: The mechanism of nanobump formation in femtosecond pulse laser nanostructuring of thin metal films. Appl Phys A 2008, 92: 791–796.

14. Vonallmen M, Lau SS, Maenpaa M, Tsaur BY: Phase-Transformations in Laser-Irradiated Au-Si Thin-Films. Appl Phys Lett 1980, 36: 205–207.

15. Zhang Y, Chen JK: Melting and resolidification of gold film irradiated by nano- to femtosecond lasers. Appl Phys A 2007, 88: 289–297.

16. Manickam S, Venkatakrishnan K, Tan B, Venkataramanan V: Study of silicon nanofibrous structure formed by femtosecond laser irradiation in air. Opt Express 2009, 17: 13869–13874.

17. Sivakumar M, Venkatakrishnan K, Tan B: Study of metallic fibrous nanoparticle aggregate produced using femtosecond laser radiation under ambient conditions. Nanotechnology 2010, 21: 225601.

18. Sivakumar M, Venkatakrishnan K, Tan B: Synthesis of Nanoscale Tips Using Femtosecond Laser Radiation under Ambient Condition. Nanoscale Res Lett 2010, 5: 438–441.

19. Sarkar DK, Dhara S, Gupta A, Nair KGM, Chaudhury S: Structural instability of the ion beam-mixed Au/Si(111) systems at

elevated temperatures. Nuclear Instrum Methods Phys Res B 2000, 168: 21–28.

20. Cullity BD, Stock SR: Elements of X-ray diffraction. 3rd edition. Upper Saddle River, NJ: Prentice Hall; 2001.

21. Khalfaoui R, Benazzouz C, Guittoum A, Tabet N, Tobbeche S: Irradiation-induced gold silicide formation and stoichiometry effects in ion beam-mixed layer. Vacuum 2006, 81: 45–48.

22. Sarkar DK, Bera S, Dhara S, Nair KGM, Narasimhan SV, Chowdhury S: XPS studies on silicide formation in ion beam irradiated Au/Si system. Appl Surf Sci 1997, 120: 159–164.

23. Sundaravel B, Sekar K, Kuri G, Satyam PV, Dev BN, Bera S, Narasimhan SV, Chakraborty P, Caccavale F: XPS and SIMS analysis of gold silicide grown on a bromine passivated Si(111) substrate. Appl Surf Sci 1999, 137: 103–112.

24. Lu ZH, Sham TK, Norton PR: Interaction of Au on Si(100) Studied by Core Level Binding-Energy Shifts. Solid State Commun 1993, 85: 957–959.

25. Ivanov PR, Zhigilei LV: Combined atomistic-continuum modeling of short-pulse laser melting and disintegration of metal films. Phys Rev B 2003, 68: 064114.

26. Ni XC, Wang CY, Yang L, Li JP, Chai L, Jia W, Zhang RB, Zhang ZG: Parametric study on femtosecond laser pulse ablation of Au films. Appl Surf Sci 253: 1616–1619.

27. Wong YY, Yahaya M, Salleh MM, Majlis BY: Controlled growth of silicon nanowires synthesized via solid-liquid-solid mechanism. Sci Technol Adv Mater 2005, 6: 330–334.

Chapter 5

Effects of Laser Pulse Numbers on Surface Biocompatibility of Titanium for Implant Fabrication

Mitra Radmanesh, Amirkianoosh Kiani

Department of Mechanical Engineering, University of New Brunswick, Fredericton, Canada

ABSTRACT

Generally, materials with high biocompatibility are more appropriate for bone and tissue transplant applications, due to their higher effectiveness in the healing process and infection problems. This study presents the effects of laser surface texturing on the surface topography properties, roughness, and wettability of thin titanium sheets, which consequently enhance the biocompatibility of this material. Creating line patterns across the surfaces, the titanium samples are prepared using variety of laser parameters. The apatite inducing ability of each sample is tested through the use of simulated body fluid (SBF). The final biocompatibility level of titanium samples is analyzed through wettability, surface angle measurements, and average surface temperature profile. Overall, the effects of laser parameter, pulse numbers, upon the biocompatibility of titanium are thoroughly examined, with results indicating that a scanning speed of 100 μm/ms results in desirable bone type apatite inducing abilities across the surface of treated titanium sheets.

Keywords: Laser Surface Micro Texturing, Titanium, Implant, Biocompatability, Hydroxyapatite Deposition

1. INTRODUCTION

The fields of biomedical and tissue engineering have been expanded immensely in previous decades. Development of new technologies to enhance the biocompatibility of a wider range of materials in a short and efficient manner is one of the important challenges in these fields. Biocompatibility issues are notably improved by the introduction of laser surface micro/nano texturing techniques, and through the fast, precise, and reliable operation of these methods. Micro/sub-micro treated titanium surfaces are popularly used in scaffold systems in bone and tissue implant applications. High level of biocompatibility of titanium, is essential in the implant fabrication for avoiding any infections and immune system rejection while providing the required structural properties for the implant [1] [2] . An effective way of enhancing the biocompatibility of a material is to alter its surface properties, such as surface structure and topography [3] . The surface topography properties of a material, espeically the roughness properties, are very influential on the apatite inducing ability of bone type apatite to surfaces. Increasing the apatite deposition rate across the surface of a material increases the biocompatibility of that material as well [4] [5] . The purpose of this paper is to investigate the effects of laser surface texturing on enhancing the biocompatibility of titanium. The surface topography properties of the titanium samples are altered through a range of laser process parameters, and effects of pulse rate on surface structure, roughness, and wettability of titanium are examined. The biocompatibility of the processed samples is then examined by performing wettability analysis and through the use of simulated body fluid (SBF) and the comparison of multiple samples. The essential purpose of this research is to study the behavior and effects of laser scanning parameters. Due to the fact that these parameters vary in a wide range, understanding their effects on surface treatment and biocompatibility could provide an efficient manner for utilizing the lasers for biomedical applications.

2. EXPERIMENTAL SETUP

2.1. Laser System & Sample Preparation

Thin sheets of titanium were chosen for conducting the experiment. The samples were initially irradiated with nanosecond pulses to create the predetermined line patterns on their surfaces with specific laser parameters. The experiment was completed using a pulse repetition rate (frequency) of 25 kHz; while the scanning speed was set to 100 μm/ms, 300 μm/ms, and 500 μm/ms, using the EZCAD© software. The laser power throughout the experiment was measured to be 7 W, which was calculated using a power-meter prior to processing of the samples. The following Table 1 displays the experiment design parameters in more detail.

2.2. SBF Processing & in Vitro Assessment

In the next phase of the study, the surface-treated samples were soaked in simulated body fluid, in order to assess their altered biocompatibility. To prepare simulated body fluid, a strict step-by-step method had to be followed. Since SBF is supersaturated with respect to apatite, failing to follow the preparation procedure accurately could lead to precipitation of apatite in the solution [6] . Given the exact amount of the required ingredients, the mixture of SBF must be conceived at controlling conditions, in terms of temperature and pH levels. After dissolving the exact amount of the given chemicals in order, using Tris and Hydrochloric acid, the pH level of the solution was controlled to be in the range of 7.40 - 7.45 at a temperature of 36.5 degrees Celsius exactly. It was essential to ensure that the pH of the solution did not increase beyond 7.45 at any point during the preparation. The solution was finalized once the pH level of the solution was adjusted at 7.40 at a temperature of 36.5 degrees Celsius exactly [6] .

Upon conceiving the simulated body fluid, the surface treated titanium sheets were soaked in the fluid for a period of three days. The soaked samples were then examined using scanning electronic microcopy and by performing surface angle measurement analysis across their surfaces. The obtained results are presented and discussed in the following sections.

Table 1. Laser parameters used in surface treating of samples.

Laser Parameters			
Frequency (KHz)	Power (W)	Scanning Speed (μm/ms)	Number of Pulses
		100	5
25	7	300	1.7
		500	1

3. RESULTS & DISCUSSION

3.1. Structure of Line Patterns across the Surface

Surface topography properties of materials are directly influential on their apatite-inducing ability. Essential laser parameters such as pulse numbers can effect these properties greatly in many ways. Generally, upon laser irradiation, the surface structure and patterns of material change result in increased roughness and surface oxidation. Surface roughness is directly related to the availability of the exposed surface of a material. The exposed surface area of a work piece is more readily available upon an increase of surface roughness, which in return enhances the apatite-inducing ability and cell adhesion rate of a material. A larger surface area is exposed, and a larger number of cells can interact with surface; hence, a larger rate of apatite inducing can take place, which is the essential goal of this experiment [5] [7] [8] . Further, laser treatment of the titanium results in the creation of thin layers of titanium oxide (titania), through an increase in surface temperature of material up to the oxidation temperature [9] [10] . The created layer of titania across the surface of titanium, results in a cushion-type layer available for cells to more easily attach across the surface; this therefore results in a higher apatite perception across the treated surface of the titanium sheets [11] . An increase in the oxidation of the surfaces is directly effective on increasing the wettability of surface materials, which is also directly related to surface roughness. This leads to an increase in the apatite-inducing ability of the material, and greatly improves the biocompatibility of the implant surfaces [9] . This effect is further analyzed through the obtained results of this study.

3.2. Surface Angle Measurements; Wettability Analysis

Surface roughness and wettability are directly related. By increasing the surface roughness and oxidation of a material, the wettability of the material increases as well, which is an indicator of the ability of that particular surface to absorb more. This results in a porous effect upon the surface of the material, and is very effective in enhancing the apatite-inducing ability of the material [11] [12] . Upon imposing the line patterns on the surfaces of titanium sheets, their wettability is analyzed. Figure 1 displays the average surface angles measured for samples prepared with scanning speeds of 100 μm/ms, 300 μm/ms, and 500 μm/ms.

Scanning speed directly influences the number of laser pulses. An increase in the scanning speed decreases the number of pulses irradiated upon the surface of the samples; hence, a scanning speed of 500 μm/ms has the lowest pulse numbers (Table 1).

As shown in Figure 1, the surface angle of the samples increases as the scanning speed increases. This means that at lower pulse numbers there is a larger contact angle between the surfaces of titanium and water. The increase in the contact angle is an indicator that with a decrease in pulse number, the surface of the titanium is modified from hydrophilic to hydrophobic. This is due to the fact that lower pulse numbers have much less effect on the surface roughness and oxidation of material, since there is a shorter period of interaction between the laser and the material's surface [11] [13] . The hydrophilic properties of a surface cause the liquid to penetrate deeper into the surface; consequently, a smaller contact angle results, which is a strong indicator of improvement in wettability of that material. Having a high wettability produces a porous temple for the cells to easily absorb and attach to the surface of titanium, and thereby enhance its biocompatibility. Looking at Figure 1, it is easy to see that the wettability of the sample prepared with a scanning speed of 100 μm/ms (pulse rate: 5) has the highest wettability due to the largest number of pulses. As mentioned, the higher number of pulses notably increases the roughness and oxidation of the titanium surface, and results in a better porous effect across the surface [7] [11] [14] .

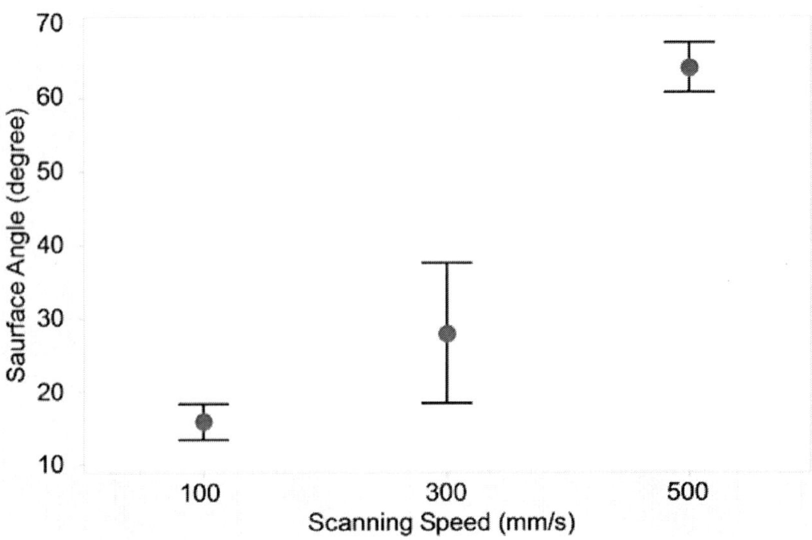

Figure 1. Average surface contact angle - wettability analysis.

3.3. SBF Soaking of Surface Treated Titanium Sheets

To fully examine the effects of pulse number on roughness, oxidation, and wettability of titanium sheets, and their biocompatibility, the samples were soaked in SBF for three days. Figure 2 and Figure 3 display the SEM micrographs of the apatite-inducing ability of samples prepared by power 7 W and scanning speeds of 100 μm/ms, 300 μm/ms, and 500 μm/ms.

As it can be seen from Figure 2, the imposed line patterns on the surface of titanium sheets has made the surface rougher, and as a result, a thicker bone-type apatite layer precipitated upon each surface. The deposition of these layers is an indicator of the apatite-inducing ability of the processed samples, and hence a higher biocompatibility. The thicker the bone-apatite layer deposition is, the higher the apatite-inducing ability is, and therefore the higher the biocompatibility of the titanium sheets is.

Figure 3 displays a closer view of the titanium samples soaked in SBF. Comparing the three samples in this figure, it can be seen that the bone like apatite layer across Figure 3(A) is more developed compared to the other two samples. As mentioned, a lower scanning speed results in a higher number of pulses. In this sample, more energy was transferred from laser to the surface of titanium, and the roughness of the surface is highly altered compared to other two cases. Therefore, a better apatite-inducing ability results in this sample. Similarly, comparing Figure 2(B) and Figure 2(C), the apatite-inducing ability of Figure 3(B) is better than the previous sample, since it has a larger number of pulses in comparison.

Further, as stated, the heat rise is larger with more number of pulses; therefore, the temperature of a larger area of the surface would reach the oxidation level. This in return will turn a larger amount of titanium into titanium oxide. As stated, titanium oxide increases the wettability of the surface material. This is in agreement with the results obtained from the wettability analysis of the titanium sheets. Sample A) with the greatest number of pulses, had the lowest contact angle, and hence the highest wettability. As stated, when the wettability of a material is increased, a higher surface energy exists across the surface of that material, and therefore a better meditation of bone-like apatite deposition can be achieved. Essentially, the higher apatite inducing ability provides a better cell attachment rate [11] [14] [15] .

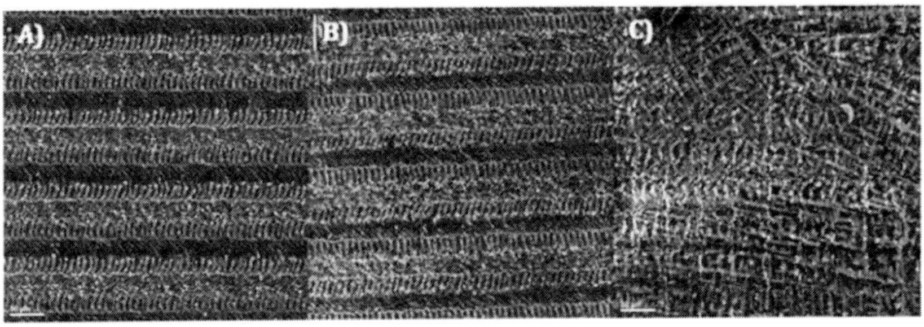

Figure 2. A) 100 μm/ms; B) 300 μm/ms; C) 500 μm/ms; Power: 7 W, F: 25 KHz; Overview of the Roughness Level of each Scanning Speed and the According Apatite Deposition Layer.

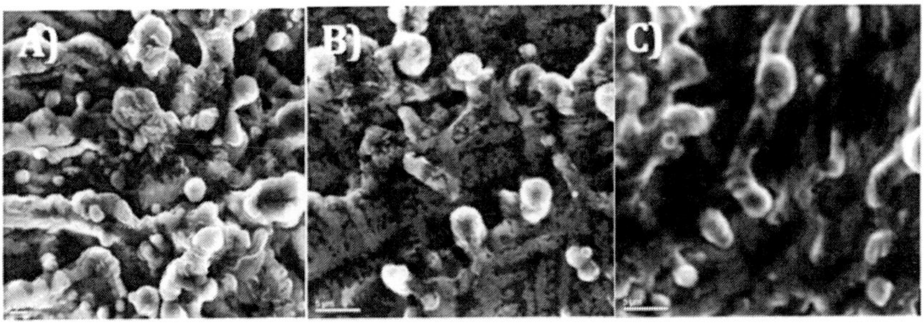

Figure 3. Zoomed SEM images of A) 100 μm/ms; B) 300 μm/ms; C) 500 μm/ms; Power: 7 W, F: 25 KHz

Figure 4 shows the magnified SEM image of the deposited ion-substituted carbonate apatite layer formed by SBF treatment. In this figure, the arrows refer to the bone-like apatite fully developed across the treated surface of the samples, much similar to the apatite layers in Figure 3.

3.4. Temperature Profile Analysis of Treated Samples

To verify the accuracy of the obtained observations, a thorough theoretical analysis of effects of laser parameters upon the surface roughness and wettability of the samples is conducted. In order to fully understand the degree of change in surface roughness and wettability of each sample, tracking the changes in the surface temperature across the samples due to number of pulses is an ideal approach. In this study, the one-dimensional heat conduction equation is assumed to be [16] :

$$\frac{\partial T}{\partial t} = a\frac{\partial^2 T}{\partial X^2} \quad . (1)$$

where a is the thermal diffusivity coefficient.

In this analysis, the thermal diffusivity coefficient is calculated using the following relation [17] [18] :

$$a = \frac{k}{\rho c_p} \quad . (2)$$

where k is the thermal conductivity coefficient, ρ is the density, and c_p is the specific heat of titanium.

To simplify the heat conduction equation, the profile of laser pulse is taken to be a rectangular profile with the step-like rise and fall; therefore the temperature profile for each laser pulse is obtained as:

$$T(X,t) = k\sqrt{\frac{a}{\pi}}\int_0^{t_p} \frac{I_{a(\tau)}}{\sqrt{t-\tau}} \exp\left\{-\frac{X^2}{2a(t-\tau)}\right\} \mathrm{d}\tau \quad (3)$$

Next the maximum temperature, occurring at the end of each laser pulse, is required to develop the surface temperature profile, which can be approximately calculated as [18] [19] :

$$T(0,t_p) = T_{\max} k \sqrt{\frac{2a}{\pi^3 t_p}} \frac{4K(1-R)P}{k\mathcal{F}d^2} \quad . (4)$$

where t_p is the pulse duration, k is the thermal conductivity coefficient, P is the average power, $\mathcal{F}$ is the frequency, d is the laser spot diameter, R is the reflection coefficient, and K is the residual energy coefficient. The reflection coefficient is calculated to be 0.56 for wavelength of 1064 nm, the residual energy coefficient is taken as 0.45, the thermal diffusivity coefficient is obtained 7.068exp-6, and laser spot diameter calculated as 30 µm.

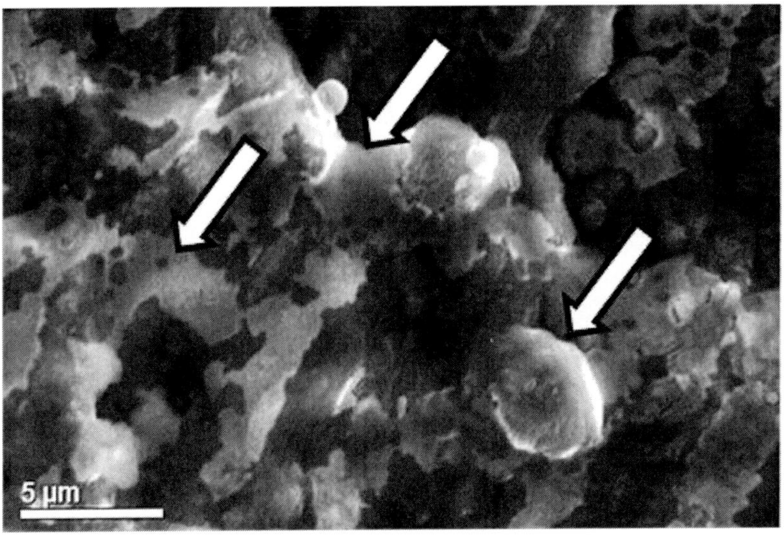

Figure 4. SEM image of deposited apatite layer after soaking in SBF; A closer magnification of the apatite developing across the surface of titanium.

Finally, the average temperature across the surface of the samples is obtained as [17] [19] :

$$T_n = 2a\frac{\left[1-\frac{2}{3}\alpha\right]}{\left(1+\alpha^2\right)}\frac{T_n}{\left(1+\alpha\right)}\left[1+\frac{\alpha^n-\alpha}{n\left(1-\alpha\right)}\right] \quad . (5)$$

where T_m is the maximum temperature calculated using Equation (1), n is the pulse numbers, and α is laser pulse which is given by $T_{min} = \alpha T_{max}$. Where α is the constant ratio for the previous maximum and the following minimum temperatures and equals to The presented analytical results of the study are associated with possible errors due to multiple assumptions made throughout the work. However, the discussed results accurately display the trend of temperatures, which is the most important aspect of the analysis. The temperature profile using power of 7 W, Frequency of 25 KHz, for each of the cases of study is obtained by using a pulse number range of 1 pulse through 50 pulses. Figure 5 displays the temperature profile obtained in this study:

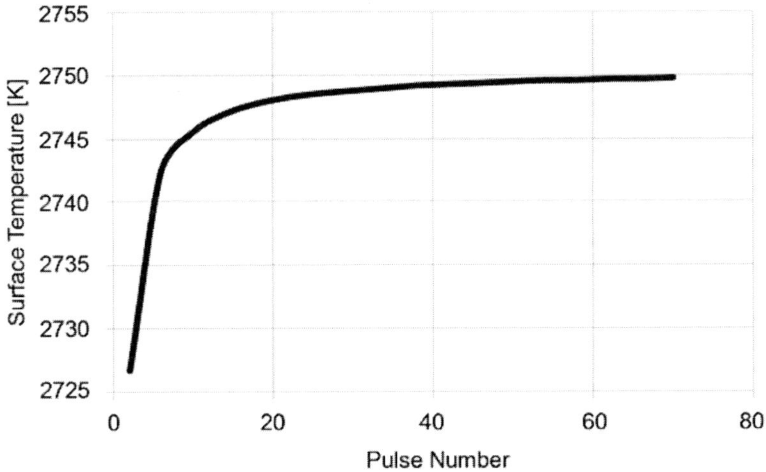

Figure 5. Average surface temperature profile based on a range of pulse numbers; Power 7 W.

Figure 5 displays the average surface temperature for the samples using a power of 7 W. As shown, the temperature rises with an increase in the number of pulses irradiated on the surface. The high temperature causes more ablation on the surface of the material, which in return increases the roughness and oxidation of the surface. The trend of temperature change is in very close agreement with the obtained observations in this study. At lower scanning speeds, a higher number of pulses are obtained, and therefore a higher surface temperature is achieved. Therefore, the sample prepared using a scanning speed of 100 µm/ms has higher roughness across the surface with a better wettability. Examining the apatite-inducing ability of this sample as discussed in Figure 2 shows this clearly [20] .

4. CONCLUSION

Through this study, the effects of laser surface texturing on the apatite-inducing ability of titanium were investigated, using different scanning speeds (thus different pulse numbers), in preparation of line patterns across the surface of the samples. By increasing the surface roughness of the samples, the wettability of the samples increased, and hence a cushion-type temple was created that produced an effective environment for cell attachment to take place. The biocompatibility of the titanium sheets was thereby enhanced to be more applicable in bone and tissue transplant applications. The samples, after pre-defined line patterns on their surfaces were generated, were placed in simulated body fluid for three days before being tested for their biocompatibility. The immersed samples were then analyzed using SEM and wettability to evaluate the bone-type apatite-inducing ability on the surface of each set of samples. Also, the effects of pulse numbers on the surface roughness, wettability, and oxidation level of the material were compared. Upon completion of the analysis, it is observed that using lower scanning speeds results in higher pulse numbers; therefore, higher surface temperatures at the irradiated spots and larger roughness and oxidation levels are achieved. This leads to better apatite-inducing ability across the surface of the titanium sheets. Overall, the experimental observations are in close agreement with the analytical results.

ACKNOWLEDGEMENTS

This research is funded by the Natural Science and Engineering Research Council of Canada (NSERC) and the New Brunswick Innovation Foundation (NBIF).

REFERENCES

1. Hutmacher, D.W. (2000) Scaffolds in Tissue Engineering Bone and Cartilage. Biomaterials, 21, 2529-2543.

2. Ahmmed, K.T., Ling, E.J.Y., Servio, P. and Kietzig, A.M. (2015) Introducing a New Optimization Tool for Femtosecond Laser-Induced Surface Texturing on Titanium, Stainless Steel, Aluminum and Copper. Optics and Lasers in Engineering, 66, 258-268.

3. Coelho, P.G., Granjeiro, J.M., Romanos, G.E., Suzuki, M., Silva, N.R., Cardaropoli, G., Thompson, V.P. and Lemons, J.E. (2009) Basic Research Methods and Current Trends of Dental Implant Surfaces. Journal of Biomedical Materials Research Part B: Applied Biomaterials, 88, 579-596.

4. Ulerich, J.P., Ionescu, L.C., Chen, J.B., Soboyejo, W.O. and Arnold, C.B. (2007) Modifications of Ti-6Al-4V Surfaces by Direct-Write Laser Machining of Linear Grooves. Proceedings of SPIE 6458, Photon Processing in Microelectronics and Photonics VI, 645819.

5. Erdoğan, M., Öktem, B., Kalaycıoğlu, H., Yavaş, S., Mukhopadhyay, P.K., Eken, K., Ilday, F.Ö., et al. (2011) Texturing

of Titanium (Ti6Al4V) Medical Implant Surfaces with MHz-Repetition-Rate Femtosecond and Picosecond Yb- Doped Fiber Lasers. Optics Express, 19, 10986-10996.

6. Kokubo, T. and Takadama, H. (2006) How Useful Is SBF in Predicting in Vivo Bone Bioactivity? Biomaterials, 27, 2907-2915.

7. Mirhosseini, N., Crouse, P.L., Schmidth, M.J.J., Li, L. and Garrod, D. (2007) Laser Surface Micro-Texturing of Ti-6Al-4V Substrates for Improved Cell Integration. Applied Surface Science, 253, 7738-7743.

8. van Tol, A.F., Tibballs, J.E., Gjerdet, N.R. and Ellison, P. (2013) Experimental Investigation of the Effect of Surface Roughness on Bone-Cement-Implant Shear Bond Strength. Journal of the Mechanical Behavior of Biomedical Materials, 28, 254-262.

9. Tavangar, A., Tan, B. and Venkatakrishnan, K. (2011) Synthesis of Bio-Functionalized Three-Dimensional Titania Nanofibrous Structures Using Femtosecond Laser Ablation. Acta Biomaterialia, 7, 2726-2732.

10. Wang, H.S., Liang, C.Y., Yang, Y. and Li, C.Y. (2010) Bioactivities of a Ti Surface Ablated with a Femtosecond Laser through SBF. Biomedical Materials, 5, Article ID: 054115.

11. Das, K., Balla, V.K., Bandyopadhyay, A. and Bose, S. (2008) Surface Modification of Laser-Processed Porous Titanium for Load-Bearing Implants. Scripta Materialia, 59, 822-825.

12. Kazemi, K. and Goldak, J.A. (2009) Numerical Simulation of Laser Full Penetration Welding. Computational Materials Science, 44, 841-849.

13. De Aza, P.N., Fernandez-Pradas, J.M. and Serra, P. (2004) In Vitro Bioactivity of Laser Ablation Pseudowollastonite Coating. Biomaterials, 25, 1983-1990.

14. Nolte, S., Momma, C., Jacobs, H., Tünnermann, A., Chichkov, B.N., Wellegehausen, B. and Welling, H. (1997) Ablation of Metals by Ultrashort Laser Pulses. Journal of the Optical Society of America B, 14, 2716-2722.

15. Fasasi, A.Y., Mwenifumbo, S., Rahbar, N., Chen, J., Li, M., Beye, A.C., Arnold, C.B. and Soboyejo, W.O. (2009) Nano-Second UV Laser Processed Micro-Grooves on Ti6Al4V for Biomedical Applications. Materials Science and Engineering: C, 29, 5-13.

16. Kiani, A., Venkatakrishnan, K., Tan, B. and Venkataramanan, V. (2011) Maskless Lithography Using Silicon Oxide Etch-Stop Layer Induced by Megahertz Repetition Femtosecond Laser Pulses. Optics Express, 19, 10834-10842.

17. Kiani, A., Venkatakrishnan, K. and Tan, B. (2010) Direct Laser Writing of Amorphous Silicon on Si-Substrate Induced by High Repetition Femtosecond Pulses. Journal of Applied Physics, 108, Article ID: 074907.

18. Kuang, J.H., Hung, T.P., Lai, K., Hsu, C.M. and Lin, A.D. (2012) The Surface Absorption Coefficient of S304L Stainless Steel by Nd: YAG Micro-Pulse Laser. Advanced Materials Research, 472, 2531-2534.

19. Venkatakrishnan, K., Stanley, P., Sivakumar, N.R., Tan, B. and Lim, L.E.N. (2003) Effect of Scanning Resolution and Fluence Fluctuation on Femtosecond Laser Ablation of Thin Films. Applied Physics A, 77, 655-658.

20. Ramsden, J.J., Allen, D.M., Stephenson, D.J., Alcock, J.R., Peggs, G.N., Fuller, G. and Goch, G. (2007) The Design and Manufacture of Biomedical Surfaces. CIRP Annals-Manufacturing Technology, 56, 687-711.

Chapter 6

Nanofabrication with Pulsed Lasers

A. V. Kabashin, Ph. Delaporte, A. Pereira, D. Grojo, R. Torres, Th. Sarnet, M. Sentis

(1) Lasers, Plasmas et Procédés Photoniques (LP3, UMR 6182 CNRS), Université de la Méditerranée
(2) Laboratoire de Physico-Chimie des Matériaux Luminescents, LPCML (UMR 5620 CNRS), Domaine Scientifique de la Doua, Université Claude Bernard Lyon 1

ABSTRACT

An overview of pulsed laser-assisted methods for nanofabrication, which are currently developed in our Institute (LP3), is presented. The methods compass a variety of possibilities for material nanostructuring offered by laser–matter interactions and imply either the nanostructuring of the laser-illuminated surface itself, as in cases of direct laser ablation or laser plasma-assisted treatment of semiconductors to form light-absorbing and light-emitting nano-architectures, as well as periodic nanoarrays, or laser-assisted production of nanoclusters and their controlled growth in gaseous or liquid medium to form nanostructured films or colloidal nanoparticles. Nanomaterials synthesized by laser-assisted methods have a variety of unique properties, not reproducible by any other route, and are of importance for photovoltaics, optoelectronics, biological sensing, imaging and therapeutics.

Keywords Laser nanofabrication Laser ablation Semiconductor and plasmonic nanostructures Colloidal nanoparticles

INTRODUCTION

When nanostructured, many materials start to exhibit new optical properties making them unique for a plethora of applications. In particular, despite small and indirect band gaps in the bulk state, the nanostructured IV group semiconductors (e.g. Si, Ge) become efficient size-dependent emitters in the visible light range [1, 2], but also can work as photosensitizers to generate singlet oxygen under photoexcitation [3, 4]. Another prominent example relates to noble metal nanostructures, which provide a number of unique plasmonic effects, including size-dependent absorption peaks [5, 6], drastic local electric field enhancement [7, 8], resolution beyond the diffraction limit [9], nanotrapping [10] etc. These new properties of emerging nanomaterials appear to be extremely promising for photovoltaics and optoelectronics, as well as for biological sensing, imaging and therapeutics.

The employment of pulsed lasers offers a novel unique tool for nanofabrication [11]. When focused on the surface of a solid target, pulsed laser radiation causes a variety of effects, including heating, melting and finally ablation of the target and such processes can lead to an efficient material nanostructuring, as shown in Fig. 1. First, the laser-assisted removal of material from the irradiated spot can result in a spontaneous formation of variety of periodic micro- and nanoarchitectures within this spot [12–16]. Second, laser ablation of material from a solid target leads to the production of nanoclusters [17–20]. When produced in gaseous environment or in vacuum, these nanoclusters can then be deposited on a substrate yielding to the formation of a nanostructured film [17, 21–24]. When produced in liquid environment, the nanoclusters can be released into the liquid forming a colloidal nanoparticle solution [25–32]. In all cases, properties of formed nanostructures can be unique and not reproducible by any other route [27–33]. As an example, the fabrication of nanoparticles in aqueous solutions does not require any chemical reducing agent, which conditions unique surface chemistry and purity of produced nanomaterials [28, 29]. Furthermore, when synthesized in clean, biocompatible environment, laser-synthesized nanomaterials are exempt of any residual toxicity that is typical for chemically synthesized nanoparticles [32, 33].

In this paper, we review laser-assisted technologies, developed by LP3 members, which are now available in our Institute.

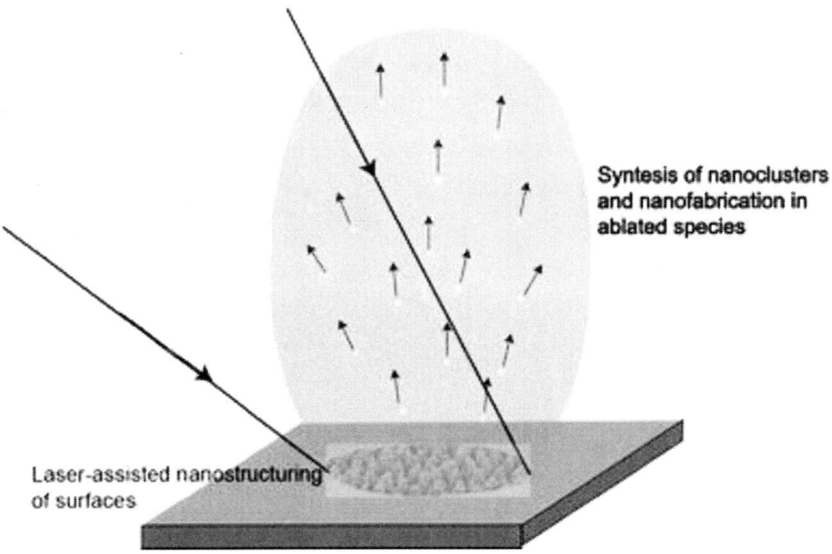

Figure 1. Schematics of laser–target interaction and material nanostructuring

LASER-ASSISTED NANOSTRUCTURING OF SURFACES

Properties of laser-materials interaction are known to be strongly dependent on parameters of laser radiation. Among these parameters, the wavelength and pulse length are especially important to determine the efficiency of radiation absorption, dynamics of plasma plume expansion and nanoclustering [11]. Ultraviolet, visible and near-infrared lasers are normally considered as most adequate for laser ablative tasks. Indeed, most industrially important materials efficiently absorb laser radiation in the spectral range of 200–1,000 nm, yielding to material ablation and nanostructuring. In addition, radiation of UV and visible lasers is relatively transparent to laser plasma, which minimizes laser beam distortions and power losses before reaching the target. In contrast, the interaction of infrared radiation (1–11 μm) with materials is characterized by a strong energy absorption by plasma itself, which completely changes conditions of nanostructuring. The pulse length is another important parameter in laser–matter interaction. Micro- and nanosecond laser–matter interactions are typically associated with long ablation regime, in which the ablation process takes place during the laser pulse itself. In contrast, pico- and femtosecond lasers offer short ablation regimes, in which moments of radiation absorption and material ablation are temporally separated. Although multi-pulse laser ablation from the target surface is always accompanied by the formation of micro- and nanoscale features and periodic structures [12–16, 34], properties of these features strongly depend on the wavelength and pulse length. Therefore, depending on the task, one can select appropriate radiation parameters to condition the ablation regime and obtain prescribed properties of laser-synthesized nanostructures. Below, we give several examples of laser-assisted methodologies to fabricate nanostructures within the irradiation spot.

Femtosecond Laser Ablation of Si: Formation of Black Si for Photovoltaics Applications

It is accepted that femtosecond pulses give at least two major advantages to micromachining compared to nanosecond and longer pulses [35, 36]: (1) the reduction of the pulse energy which is necessary to induce ablation for fixed laser wavelength and focussing conditions and (2) a significant reduction or complete removal of heat-affected zone (HAZ) and, as consequence, the improvement of the contour sharpness for the laser-processed structures. The second advantage is a direct consequence of the pulse being shorter than the heat diffusion time, given by the phonon transport. As in the case of nanosecond- and microsecond laser ablation, multi-pulse femtosecond ablation leads to a spontaneous formation of nanoarchitectures [14, 37–40]. However, in contrast to long ablation, the fs regime is characterized by the absence of target melting effects, yielding to the formation of clean micro- and nanoscale features. In particular, using multi-pulse fs ablation of Si in the presence of SF_6 reactive gas, Mazur et al. [14, 37] managed to fabricate extremely narrow micro-spikes within the irradiation spot, which are capable of efficiently absorbing light in the visible and infrared ranges. The efficient absorption of light exceeding 95% in the visible was attributed to a geometric multi-reflection effect offered by a unique spike-based structure, while the enhanced absorption in the infrared was explained by a sulphur doping [37]. Due to the wide-range absorption effect, the produced spike-based structure was called "black silicon" and was later used for the development of Si-based photodetectors. It is worth noting that such absorptive features cannot be reproduced by any alternative non-laser route. Other studies (see, e.g. [41]) reported the fabrication of nanostructured metal films exhibiting colours ("coloured metals") using similar fs ablation approach.

Our sub-project is devoted to the fabrication of "black silicon" structures for photovoltaics solar panel applications. The choice of photovoltaics as target application imposes new criteria on nanostructuring conditions. First, these applications require a high absorption mostly in the visible—near-IR range (300–1,000 nm), which enables us to exclude the necessity of using sulphur-based doping species. Second, these applications require uniform high-quality doping of nanostructured layers to maximize the photovoltage response. We succeeded in developing of a novel methodology to produce "black silicon" with such parameters, employing a Ti:Sapphire laser (wavelength 800 nm, pulse energy 5 mJ, repetition rate 1 kHz) [42]. In contrast to [14, 37], we carry out multi-pulse laser processing in vacuum under the residual pressure of $(1–5) \times 10^{-5}$ mBar. In addition, we avoid the doping procedure during the laser processing process and do it

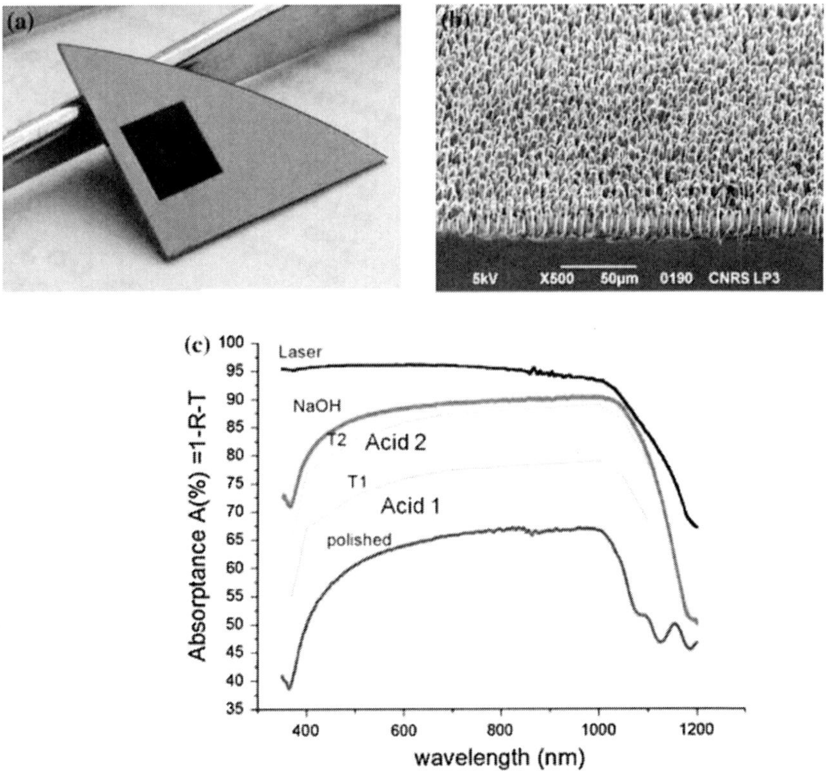

Figure 2. a Typical image of "black silicon" spot fabricated on a Si wafer by multi-pulse fs laser ablation in vacuum; **b** Typical scanning electron microscopy (*SEM*) image of penguin-like structure of black silicon; **c** Typical absorption spectra from "black silicon" and silicon treated by different methods

afterwards. To achieve a high quality doping of deep layers, the laser-structured samples are boron implanted by Plasma immersion (PULSION, BF3, 2 kV, 900°C, 30 min) and thermally annealed (TA). The junction depth obtained by this method is estimated to be about 150 nm, which is much shallower than the 3D laser structures; therefore, the junction follows the topography of the structures. Figure 2a demonstrates a silicon wafer surface after the laser processing and boron implantation procedure. Here, a rectangular area of 3 × 2 cm^2 is written by a programmed displacement with the speed of 150 μm/s of a femtosecond laser beam having the spot size of 35 × 35 μm^2. One can clearly see a black area on the silicon wafer associated with "black silicon". As shown in Fig. 2b, the treated surface contains penguin-like nanospikes with the length of up to 10 μm and sub-μm lateral dimensions. Although the morphology of femtosecond laser-treated surface is rather different compared to narrow spike-like structures in [14, 37], it is also characterized by an enhanced absorption in the visible range exceeding 90% (Fig. 2c). Depositing grating-like contacts on the top on the treated area, we were able to obtain the amplification of photocurrent by 50% compared to the untreated surface area. Such result was attributed to an enhanced absorption granted by the penguin-like structures, much larger surface of nanostructured silicon used for signal collection, and high quality of boron implantation offered by the post-ablation plasma implantation procedure. The fabricated structures are now actively tested as photovoltaics solar cells.

Laser Plasma-assisted Nanostructuring of Surfaces

As we mentioned above, UV or ultrashort lasers contribute to a good radiation absorption by the target itself, while plasma remains relatively transparent to the incoming radiation. Such parameters ensure good quality of surface treatment in laser processing tasks. The plasma effect can be further minimized by reducing the pressure of the ambient gas. Depending on plasma plume size conditioned by the ambient gas pressure, the material can be re-deposited either within the irradiation spot (for high atmospheric pressure) or into the environment (for reduced pressures). In particular, for atmospheric pressure, the ablation process results in the formation of a deep crater, containing microscale spikes, covered by re-deposited nanoparticles [13–16]. In this case, chemical transformations in ablated species are minimal, since the ablated material rapidly cools down while interacting with the environment [24, 43].

We recently introduced a novel method for surface nanostructuring, which is characterized by radically different nanofabrication conditions [44–49]. The method may look paradoxical, since it disaccords with main principles of laser processing requiring the minimization of plasma-related effects as one of main conditions to achieve high quality of laser treatment. In contrast, in this method, plasma-associated effects are amplified by all possible means. Basically, we use infrared radiation from CO_2 laser, which is strongly absorbed by the plasma itself. When focused in air and any other gas having atmospheric pressure, infrared radiation is capable of efficiently igniting the gas breakdown and this phenomenon is called the "laser spark". The presence of a target decreases the breakdown threshold by 2–3 orders of magnitude [50]. In the latter case, the target serves to provide first electrons. Then, an avalanche plasma discharge develops in ambient air moving towards the focusing lens. Absorbing main radiation power through the inverse Bremsstrahlung mechanism, the plasma accumulates an enormous amount of energy and is supposed to radically change conditions of nanocluster production and growth [51]. Indeed, in contrast to conventional laser ablation, the ablated species find themselves in a plasma "reactor" with extremely high temperatures (10^4 K) [52] and strong electromagnetic fields [53–56], yielding to a deep chemical transformation of properties of ablated clusters. The clusters then move back to the irradiated spot forming a film of clearly separated and densely packed spherical nanoparticles, as shown in Fig. 3a. The size of nanoparticles can vary for different materials, but is usually between 20 and 70 nm. The increase of plasma intensity can also lead to a coagulation of nanoparticles and the formation of much larger microscale spherical features. Nanostructures treated by this method have a specific texture with separated densely packed crystalline nanoparticle constituents, which contribute to unique optical properties. In particular, in the case of the laser plasma-based treatment of Zn in ambient air, the produced ZnO nanostructures exhibit very strong exciton-related peak around 380–385 nm under photoexcitation, whereas photoluminescence peaks associated with defects are essentially absent [49]. Furthermore, such nanostructure is capable of providing the mirror-less random lasing effect, arising as a result of a simultaneous strong amplification and scattering in a highly disordered medium [48]. Such effect is normally observed by the appearance of several extremely narrow lines within the exciton emission band under the increase of the pumping laser power. In the case of Si and Ge, the laser plasma treatment leads the formation of nanostructures, which are capable of generating strong photoluminescence (PL) in the visible [44–47]. Figure 3b shows PL properties of Si nanostructures fabricated by the laser plasma-based treatment. One can see two PL bands around 2.1 and 3.25 eV, associated with Si-based nanostructures. In the case of Ge, the PL bands are slightly different and situated around 2.2 and 2.9 eV [47].

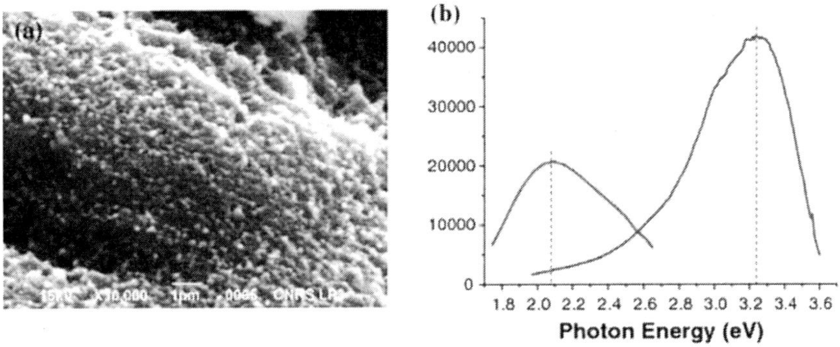

Figure 3. **a** Typical image of Si-based nanostructures prepared by laser plasma-assisted treatment of a Si wafer; **b** Photoluminescence spectra from laser plasma-treated nanostructured Si spots for different pumping wavelengths (325 and 488 nm)

Near-field Nanoparticle-assisted Nanostructuring of Surfaces: Fabrication of Patterned Nanoarrays

This sub-project addresses the formation of periodic nanoarrays by laser-assisted methodologies. The methodology implies two steps [57]: (1) laser ablation to form programmed periodic nanoscale features on a

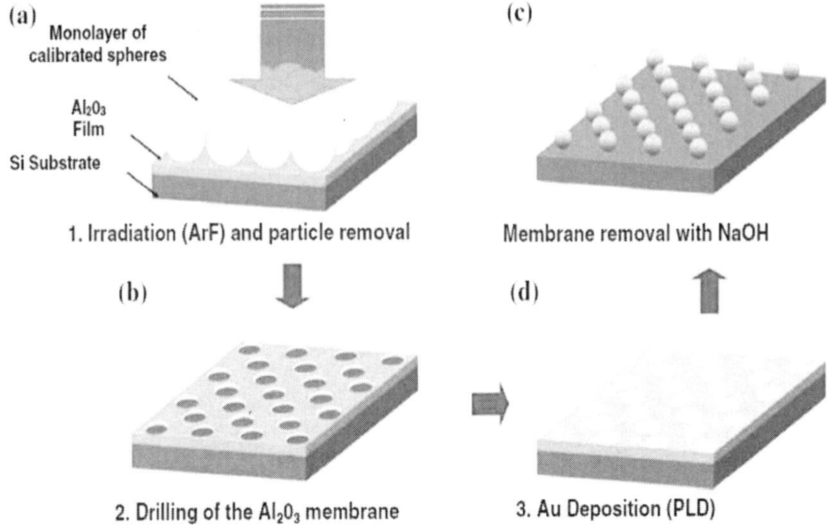

Figure 4. Nanodot array fabrication. **a** A monolayer of spheres deposited on a thin alumina film is illuminated with a single laser pulse. **b** Near-field enhancement underneath the spheres leads to the parallel nanodrilling of the film. **c** A metal (gold in our case) is then deposited and the alumina membrane is dissolved in basic solution. **d** An ordered Au nanodot array is then obtained on the silicon substrate

sacrificial surface layer; (2) post-ablation deposition/chemical treatment step to fabricate nanoarrays. The first step is based on the use of near-field particle-assisted ablation to produce nanoscale features on various substrates [58–61]. A monolayer of self-assembled SiO_2 spheres is formed on a 20-nm alumina (Al_2O_3) film, as shown schematically in Fig. 4a. Then, pores are optically drilled in the Al_2O_3 film by particle-assisted near-field enhancement. This is accomplished through the illumination of the spheres with a single nanosecond laser pulse at the wavelength of 193 nm. Such process leads to the local removal of the 20-nm-thick Al_2O_3 film under each sphere. Since the spheres are arranged in a hexagonal array at the surface of the substrate, the Al_2O_3 film is decorated with an ordered arrangement of holes (Fig. 4b). The second step employs the laser-fabricated porous alumina membrane (LF-PAM) as a mask for the deposition of metal (Fig. 4c). Then, the alumina layer is dissolved yielding to the production of a series of ordered metal nanodots on the surface of the substrate (Fig. 4d). One of the main advantages of the proposed method consists in a possibility of producing nanodot arrays of functional materials, independently of the nature of the substrate.

In particular, the proposed methodology can be used to fabricate an ordered array of gold nanodots (plasmonic arrays). Figure 5a shows an image of an Al_2O_3 thin film after drilling holes by 250-nm silica particle-assisted laser ablation. One can see that the laser-drilled holes are relatively uniform with the mean size of holes of about 100 nm, while the distance between the nanoholes is well conditioned by the size of self-assembled silica microparticles. As shown in Fig. 5b, 5c, the second deposition/chemical treatment step leads to the formation of high quality plasmonic arrays based on gold nanodots with the size of features less than 100 nm. These nanodot arrays are of importance for biosensing applications [62, 63].

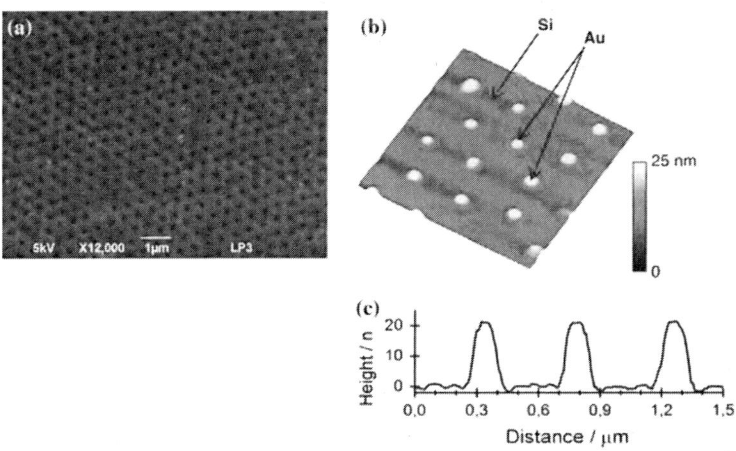

Figure 5 **a** SEM image of an Al_2O_3 thin film (20 nm) deposited on a Si substrate and simultaneously drilled by the near-field enhancement of a single nanosecond laser pulse, which is produced by a lattice of SiO_2 spheres (R = 250 nm); **b** AFM image and **c** depth profile of gold nanodots created on silicon substrates by the LF-PAM-based process

LASER ABLATIVE SYNTHESIS OF NANOCLUSTERS

The evolution of properties of formed nanostructures after the ablation process is mainly determined by the interaction of the species with the environment. In fact, in the first approximation, the nanocluster formation process can be described by the classical theory of condensation and nucleation in a vapour layer [64, 65]. However, the growth of nanoclusters strongly depends on initial parameters of ablated species (energy, angular distribution, density) and by laser interaction with the environment, which are conditioned by characteristics of pumping radiation. In particular, the radiation can directly affect the nucleation process if the pumping laser pulse is long enough. Indeed, the energy of photons can be sufficient to produce nucleation centres, change the dynamics of the nuclei growth and modify the diffusion of species in the vapour phase. Below, we give a review of experimental results on the laser ablative nanostructure growth.

Laser Ablation in Residual Gases and Deposition of Nanostructured Films

The first possibility is related to the ablation of material in gaseous environment and the deposition of ablated species onto a substrate in pulsed laser deposition (PLD) geometry. In this geometry, the employment of a UV radiation from an excimer laser is normally preferable [11]. UV radiation is well absorbed by most industrially important materials, while the formed plasma plume is relatively transparent to it. The material can be in general ablated in vacuum; but due to a low probability of nanocluster coalescence in vacuum, it is difficult to control their growth and the nanoclusters normally deposit as ablated, forming a dense film with a significant amorphous fraction. Therefore, it is important to have a residual neutral light gas to affect the growth procedure. In this case, nanoclusters cool down under collisions with gas molecules or atoms, which contribute to their coalescence in the vapour phase. Under such conditions, the nanocluster growth process can be efficiently controlled by varying the pressure of the ambient gas [23, 66, 67].

To fabricate Si-based nanostructured films, we normally use radiation of a pulsed ArF or KrF lasers (193 or 248 nm, respectively, 15 ns FWHM, repetition rate 30 Hz) to ablate material from a rotating Si target. The radiation is focused at the incident angle of 45° to the surface. A substrate is placed on a rotating holder in front of the target. The experimental chamber is filled with helium for a deposition at a constant pressure ranging between 0.05 and 10 Torr. The film thickness after several thousands laser shots is 100–700 nm.

Figure 6 shows a transmission electron microscopy (TEM) image of several isolated nanoparticles, synthesized by laser ablation from a Si target and deposited on a carbon-coated TEM grid (a) and corresponding nanoparticle size distribution (b). One can see that the produced nanoclusters are very small with the size in the range of 1–4 nm. As shown in Fig. 6c, the mean size of nanoparticles slightly increases under the increase of He pressure. For example, the increase of the pressure from 0.5 to 8 Torr results in the increase of the nanoparticle size from 1.5 to 4 nm. Another important feature relates to essentially porous texture of the films prepared by pulsed laser deposition, as illustrated by the inset of Fig. 6c, and the porosity of films increases with the increase of the ambient gas pressure (Fig. 6c). Indeed, while the deposition under 1 Torr results only in some germs of roughness, the experiment under 2 Torr provides a developed porous structure with pore size of about 50–100 nm. A further pressure increase up to 4 Torr leads to a formation of web-like aggregations of particles. As shown in Fig. 6c, the deposition of films at high pressures (>4 Torr) leads to porosities exceeding 90%, corresponding to the formation of powders on the substrate. Thus, the pressure of the ambient gas appears to be one of key parameters, which determines both the size of synthesized nanocrystals and the porosity of deposited nanostructured layers. Such a strong impact of the gas pressure on nanoclustering process suggests the importance of cooling of ablated species under their collisions with gas atoms. If the pressure of the gas is high enough, the nanoclusters experience a sufficient number of collisions to rapidly cool down and crystallize in the vapour phase. As a result, they arrive on the substrate in the form of a powder. In contrast, low collision regime at low pressures advantages the formation of dense and low-porous films.

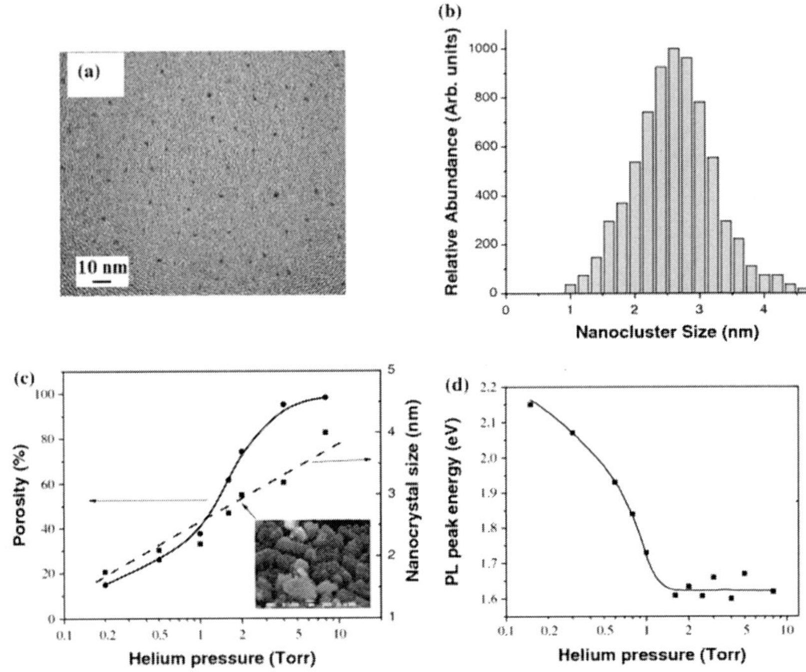

Figure 6. Transmission electron microscopy image of Si nanoparticles fabricated by pulsed laser ablation at 2 Torr of He (**a**) and corresponding nanocluster size distribution (**b**); **c** Dependence of the nanocrystal size (*dashed line*) and film porosity (*solid line*) on the pressure of He during the deposition; *Inset* typical scanning electron microscopy image of films prepared by pulsed laser ablation; **d** Dependence of the position of PL peak from laser-ablated films on the gas pressure during the deposition

It is important that all Si-based films prepared by the PLD method exhibit strong visible emission (PL) under photoexcitation, while PL properties of the films are quite different for films of different porosities. Low-porous films $P < 40\%$, deposited at reduced pressures $P < 1.5$ Torr, exhibit relatively weak PL with peak energy strongly depending on the gas pressure. In particular, the pressure decrease from 1.5 to 0.15 Torr in different depositions causes a blue shift of the peak from 1.6 to 2.15 eV, as shown in Fig. 6d. In contrast, films with an enhanced porosity $P > 40\%$, deposited at higher pressures, provide only spectra with fixed peaks. The first band (1.6–1.7 eV) is independent of the pressure and can be seen just after the fabrication process. An additional 2.2–2.3 eV band can appear under the oxidation of samples in humid air. Here, PL properties of low-porous films (porosity <40%) are almost unchangeable under these conditions, while the integral PL intensity from highly porous films significantly increases with the prolonged oxidation. We attribute such a difference of PL properties to the impact of post-fabrication natural oxidation, controlled by the level of porosity. Dense and self-coagulated structures of the films fabricated under $P < 1$ Torr minimize the impact of ambient atmosphere on the film properties; and for these films, mechanisms related to core silicon crystals became predominant. Since the blue shift of the spectra under the decrease of helium residual pressure is accompanied by a certain decrease of the nanocrystal size, the quantum confinement mechanism [1] can be considered as one of main opportunities. In contrast, a high porosity enhances the surface area, which is subjected to surface chemistry modifications due to interactions of nanocrystallites with oxygen. This can drastically enhance the role of oxidation in the formation of PL centres and the domination of oxygen-related PL mechanisms connected either to defects in the SiO_2 structure (usually, this mechanism provides 2–2.4 eV PL [68]) or to the interfacial layer (1.65 eV) [69].

Ultra-short Laser Ablation in Liquid Environment to Form Colloidal Nanoparticles

In the case of liquid ambience, laser ablation process leads to the release of nanoclusters into the liquid and the formation of a colloidal nanoparticle solution [11]. In contrast to conventional chemical reduction methods, this method enables avoidance of the use of toxic chemical reduction agents to control the growth of the nano-particles. As an example, the laser ablation-based synthesis can be implemented in pure deionized water. The independence of laser-based synthesis of dirty colloidal chemistry makes it unique for the fabrication of markers of bioanalytes for sensing and imaging applications. In pure water, however, or any other aqueous solution exempted of additional chemically active components, the size of nanoparticles produced tends to be relatively large, since a natural coalescence of hot ablated nanoclusters cannot be easily overcome. In particular, nanosecond laser-based ablation used in most works, generally gives relatively large (10–300 nm) and strongly dispersed (50–300 nm) particles [25–28]. Although certain size control can be achieved by decreasing the wavelength of pumping radiation or varying the laser fluence, the range of size variations stays rather moderate in the nanosecond pulse case. Mafune [27] showed that size characteristics of nanoparticles can be improved by adding some reactive surfactants such as sodium dodecyl sulphate (SDS) or CTAB during the ablation. As an example, thiol-containing SDS covered gold nanoclusters just after their production and thus protected them from further coalescence. However, bioimaging applications of so produced nanoparticles are hardly possible since the surfactants are not biocompatible.

We recently proposed a femtosecond laser ablation-based method for nanoparticle synthesis, which makes possible an efficient control of size of prepared nanoparticles by varying physical parameters [28–33]. The experiments are normally carried out with an Ytterbium (400 fs FWHM, 1,025 nm, 1 kHz) or a Ti/Sapphire laser (110 fs FWHM, 800 nm, 1 μJ/pulse, 1 kHz). The radiation is focused onto a target of different materials (Au, Ag, Si, Ti, Cu), which is placed on the bottom of a glass vessel filled with aqueous solutions. The vessel is placed on a horizontal moving platform to avoid the ablation of material from the same area. The ablation experiments are carried out in pure deionized water and in aqueous solutions (biopolymers, cyclodextrins).

When performed in relatively neutral environment such as deionized water, fs laser ablation normally leads to the formation of two nanoparticle populations, independently on the material of the target, as shown in Fig. 7. The first population is characterized by a small mean nanoparticle size and narrow size dispersion, whereas the second one has a much larger mean size and broader size dispersion. The presence of the two populations suggests the involvement of two different mechanisms of nanoparticle growth. The production of the first, less dispersed population is characterized by the absence of target melting effects, suggesting direct radiation-related ablation of material [28]. In contrast, the production of the highly dispersed population is accompanied by a strong melting of material inside the ablated crater. This melting is usually attributed to the explosion of a cavitation bubble formed as a result of energy transfer from laser plasma to the liquid [11]. It is important that in the fs laser ablation regime, the nanoparticle size can be efficiently controlled by varying the intensity of radiation during the nanosynthesis process. In particular, the mean size of gold nanoparticles can be reduced from 120 to 4 nm by the decrease of laser fluence down to the threshold values [28]. Similar effect can be achieved by changing the radiation focusing on the target surface [29]. Such efficient method of size control, not possible with nanosecond or microsecond pulses, is granted by specific conditions of fs laser–materials interaction. In the fs ablation regime, much less radiation energy is transferred to the cavitation bubble (15% compared to 80% in the nanosecond pulse regime [70]). We believe that the decrease of the laser fluence down to near-threshold values enables one to completely avoid cavitation phenomena and thus eliminate the second highly size-dispersed nanoparticle population. In addition, the decrease of the laser fluence changes parameters of ablated nanoclusters (energy, angular distribution etc.), which can in turn affect the final size of nanoparticles of the first population. In particular, using this method in the case of gold, we managed to vary the size of synthesized nanoparticles between 4 and 20 nm [28].

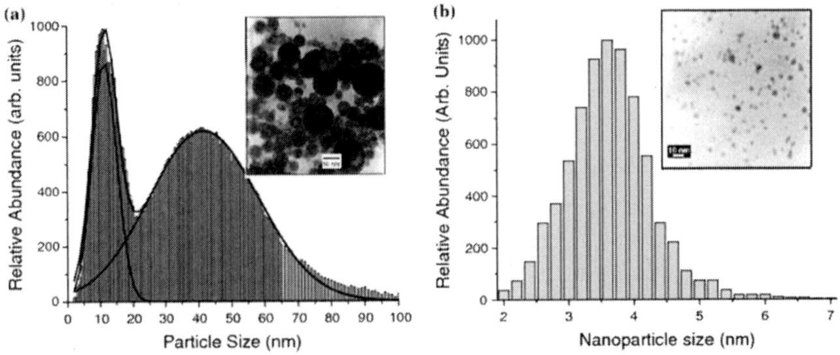

Figure 7. Transmission electron microscopy images and corresponding size distributions of **a** TiO_x nanoparticles prepared by fs laser ablation from a Ti target in deionized water; **b** gold nanoparticles prepared in aqueous solution of polyethylene glycol

Another important issue is related to chemical properties of laser-synthesized nanoparticles. Since the nanoparticles are produced by pure physical ablation from a target without the involvement of any specific chemicals, surface chemistry of these nanoparticles can drastically differ from that of counterparts prepared by conventional colloidal chemistry [30, 31, 71–73]. In particular, laser-synthesized gold becomes susceptible to oxidation and, in contrast to chemically prepared gold, the surface of these nanoparticles is partially covered by a layer of oxide. Furthermore, the oxidized nanoparticle surface can have different termination, depending on the pH of the environment. The oxidized portion of the gold surface normally has Au–O– groups at pH > 5.8 and increasing numbers of Au–OH groups at pH < 5.8 [71]. This oxide-related surface termination makes possible interactions of gold nanoparticles with groups, for which conventional gold is normally inert. A prominent example of new gold chemistry is related to the use of biopolymers [72] and oligosaccharides [30]. Although these substances do not contain gold-reactive thiol group, they react with laser-synthesized gold nanoparticles, yielding to a drastic reduction of the nanoparticle size. Figure 7b illustrates the effect of the reduction of nanoparticle mean size under the use of a biopolymer polyethylene glycol (PEG). One can see that the nanoparticles' mean size and size dispersion can be reduced down to 3 nm with the size dispersion not exceeding 1.5 nm FWHM. We believe that such reduction of nanoparticle size is a result of the hydrogen bonding of the –OH groups of these compounds and the $-O^-$ at the gold surface. The molecules of PEG cover gold nanoclusters just after ablation and act like "bumpers", limiting contact between particles, preventing their coalescence (when the particles are still "hot") and aggregation (when the particles are "cold"). Similar mechanism takes place during laser ablation in aqueous solutions of other polymers (dextran etc.) [72] and oligosaccharides (cyclodextrin) [30]. Thus, OH groups of different biocompatible compounds can efficiently react with oxidized gold surface leading to the reduction of the nanoparticle size. It is important that in contrast to SDS or other surfactants previously used to control the nanoparticle size [27], biopolymers and oligosaccharides are essentially biocompatible. Moreover, the ultra-pure laser-ablated nanoparticles can be functionalized by a proper chemical modification of chemicals. We believe that this gives a huge advantage over the chemically produced nanoparticles for nano-engineering and functionalization of nanoparticles produced, as well as for a solution of toxicity problems. In particular, PEG is known as one of best materials to avoid the immune response in "in vivo" applications of inorganic nanoparticles. When covered by PEG, the nanoparticles become invisible for the immune system. In the case of laser-synthesized nanoparticles, one does not need to use any intermediate chemical group to link nanoparticles to PEG, as it takes place in the case of chemically synthesized nanoparticles.

CONCLUSIONS

We reviewed on the development, in our Institute, of various laser-assisted methodologies for nanofabrication in the gaseous and liquid environment. The methodologies imply the fabrication of nanoparticles/nanostructures either within the laser irradiation spot on the target surface or in the ablated species. Laser-synthesized nanomaterials exhibit unique optical properties and are exempt of toxicity, which make them very important for photovoltaics, optoelectronics, biological sensing, imaging and therapeutics.

NOTES

Presented by A Pereira

ACKNOWLEDGMENTS

The authors are grateful to Agence Nationale de Recherche (ANR) and Ion Beam Services (IBS company) for Plasma Immersion doping of the black silicon.

OPEN ACCESS

REFERENCES

1. Canham LT: Appl. Phys. Lett.. 1990, **57:** 1046. COI number [1:CAS:528:DyaK3cXmt1ahurc%3D]; Bibcode number [1990ApPhL..57.1046C] COI number [1:CAS:528:DyaK3cX mt1ahurc%3D]; Bibcode number [1990ApPhL..57.1046C]

2. Cullis AG, Canham LT, Calcott PD: J. Appl. Phys.. 1997, **82:** 909. COI number [1:CAS:528:DyaK2sXkvFCntbg%3D]; Bibcode number [1997JAP....82..909C] COI number [1:CAS:528:DyaK2s XkvFCntbg%3D]; Bibcode number [1997JAP....82..909C]

3. Kovalev D, Gross E, Künzner N, Koch F, Timoshenko VY, Fujii M: Phys. Rev. Lett.. 2002, **8913:** 137401. Bibcode number [2002PhRvL..89m7401K] Bibcode number [2002PhRvL. .89m7401K]

4. Timoshenko VY, Kudryavtsev AA, Osminkina LA, Vorontsov AS, Ryabchikov YV, Belogorkhov IA, Kovalev D, Kashkarov PK: JETP Lett.. 2006, **83:** 423. COI number [1:CAS:528:DC %2BD28XmslSnurs%3D] COI number [1:CAS:528: DC% 2BD28X mslSnurs%3D]

5. Kerker M: The Scattering of Light and Other Electromagnetic Radiation. Academic Press, New York; 1969.

6. Kreibig U, Vollmer M: Optical Properties of Metal Clusters. Springer, Berlin; 1996.

7. Nie S, Emory SR: Science. 1997, **275:** 1102–1106. COI number [1:CAS:528:DyaK2sXhtlGlsL4%3D] COI number [1:CAS:528 :DyaK2sX htlGlsL4%3D]

8. Li K, Li X, Stockman M, Bergman D: Phys. Rev. B. 2005, **71:** 115409. Bibcode number [2005PhRvB..71k5409L] Bibcode number [2005PhRvB..71k5409L]

9. Yao J, Liu Z, Liu Y, Wang Y, Sun C, Bartal G, Stacy A, Zhang X: Science. 2008, **321:** 930. COI number [1:CAS:528: DC%2BD1cXpslWrtLo%3D]; Bibcode number [2008Sci...321..930Y] COI number [1:CAS:528:DC% 2BD1cXpsl WrtLo%3D]; Bibcode number [2008Sci...321..930Y]

10. Grigorenko AN, Roberts NW, Dickinson MR, Zhang Y: Nat. Photon.. 2008, **2:** 365–368. COI number [1:CAS:528:DC%2BD1cXnsFKnsbo%3D]; Bibcode number [2008NaPho...2..365G] COI number [1:CAS:528:DC%2BD1cXnsFKnsbo%3D]; Bibcode number [2008NaPho...2..365G]

11. Kabashin AV, Meunier M: **Recent advances.** In Laser Processing Material. Edited by: Perriere J, Millon E, Fogarassi E. Elsevier, Amsterdam; 2006:1–36.

12. Prokhorov AM, Konov VI, Ursu I, Mikhailescu IN: Laser Heating of Metals. Hilger, Bristol; 1990.

13. Krajnovich DJ, Vazquez JE: J. Appl. Phys.. 1993, **73:** 3001. COI number [1:CAS:528:DyaK3sXhvFalsrY%3D]; Bibcode number [1993JAP....73.3001K] COI number [1:CAS:528:DyaK3sXhv FalsrY%3D]; Bibcode number [1993JAP....73.3001K]

14. Her T-H, Finlay RJ, Wu C, Deliwala S, Mazur E: Appl. Phys. Lett.. 1998, **73:** 1673. COI number [1:CAS:528:DyaK1cXmtValtr4%3D]; Bibcode number [1998ApPhL..73.1673H] COI number [1:CAS:528:DyaK1cXmtValtr4%3D]; Bibcode number [1998ApPhL..73.1673H]

15. Pedraza AJ, Fowlkes JD, Lowndes DH: Appl. Phys. Lett.. 1999, **74:** 2322. COI number [1:CAS:528:DyaK1MXit1yiurk%3D]; Bibcode number [1999ApPhL..74.2322P] COI number [1:CAS:528:DyaK1 MXit1yiurk%3D]; Bibcode number [1999ApPhL..74.2322P]

16. Costache F, Henyk M, Reif J: Appl. Surf. Sci.. 2002, **186:** 352. COI number [1:CAS:528:DC%2BD38XitFart78%3D]; Bibcode number [2002ApSS..186..352C] COI number [1:CAS:528:DC%2B D38XitFart78%3D]; Bibcode number [2002ApSS..186..352C]

17. Movtchan IA, Marine W, Dreyfus RW, Le HC, Sentis M, Autric M: Appl. Surf. Sci.. 1996, **96–98:** 251.

18. Geohegan DB, Puretzky AA, Duscher G, Pennycook SJ: Appl. Phys. Lett.. 1998, **73:** 438. COI number [1:CAS:528:DyaK1cXksleisLg%3D]; Bibcode number [1998ApPhL..73..438G] COI number [1:CAS:528: DyaK1cX ksleisLg%3D]; Bibcode number [1998A pPhL..73..438G]

19. Geohegan DB, Puretzky AA, Duscher G, Pennycook SJ: Appl. Phys. Lett.. 1998, **72:** 2987. COI number [1:CAS:528:DyaK 1cXjsVOhurY%3D]; Bibcode number [1998ApPhL..72.2987G] COI number [1:CAS:528:DyaK1cXjsVOhurY%3D]; Bibcode number [1998ApPhL..72.2987G]

20. Povarnitsyn ME, Itina TE, Sentis M, Khishenko KV, Levashov PR: Phys. Rev. B Phys. B. 2007, **75:** 235414. Bibcode number [2007PhRvB..75w5414P] Bibcode number [2007PhRvB. .75w5414P]

21. Werwa E, Seraphin AA, Chiu LA, Zhou C, Kolenbrander KD: Appl. Phys. Lett.. 1994, **64:** 1821. COI number [1:CAS:528:DyaK2cXjtFajtbc%3D]; Bibcode number [1994ApPhL..64.1821W] COI number [1:CAS:528:DyaK2c XjtFajtbc%3D]; Bibcode number [1994ApPhL. .64.1821W]

22. Yamada Y, Orii T, Umezu I, Takeyama Sh, Yoshida T: Jpn. J. Appl. Phys.. 1996, **35:** 1361. COI number [1:CAS:528:DyaK28XisFSqu70%3D]; Bibcode number [1996JaJAP..35.1361Y] COI number [1:CAS:528:DyaK 28XisFSqu70%3D]; Bibcode number [1996JaJAP..35.1361Y]

23. Kabashin AV, Sylvestre J-P, Patskovsky S, Meunier M: J. Appl. Phys.. 2002, **91:** 3248. COI number [1:CAS:528:DC%2BD38Xhs1agsr8%3D]; Bibcode number [2002JAP....91.3248K] COI number [1:CAS:528:DC% 2BD38Xhs1agsr8%3D]; Bibcode number [2002JAP....91.3248K]

24. Pereira A, Cros A, Delaporte Ph, Georgiou S, Manousaki A, Marine W, Sentis M: Appl. Phys. A. 2004, **79:** 1433. COI number [1:CAS:528:DC%2BD2cXlslCiurw%3D]; Bibcode number [2004ApPhA..79.1433P] COI number [1:CAS:528:DC% 2BD2cXlslCiurw%3D]; Bibcode number [2004ApPhA..79.1433P]

25. Henglein AJ: Phys. Chem.. 1993, **97:** 5457. COI number [1:CAS:528:DyaK3sXisVChurc%3D] COI number [1:CAS:528: DyaK3sXisVChurc%3D]

26. Nedderson J, Chumanov G, Cotton TM: Appl. Spectrosc.. 1993, **47:** 1959. Bibcode number [1993ApSpe..47.1959N] Bibcode number [1993ApSpe..47.1959N]

27. Mafuné F, Kohno J-Y, Takeda Y, Kondow T, Sawabe H: J. Phys. Chem. B. 2000, **104:** 8333.

28. Kabashin AV, Meunier M: J. Appl. Phys.. 2003, **94:** 7941. COI number [1:CAS:528:DC%2BD3sXps1Wltbk%3D]; Bibcode number [2003JAP....94.7941K] COI number [1:CAS:528:DC%2BD3sXps1Wltbk%3D]; Bibcode number [2003JAP....94.7941K]

29. Sylvestre J-P, Kabashin AV, Sacher E, Meunier M: Appl. Phys. A. 2004, **80:** 753. Bibcode number [2005ApPhA..80..753S] Bibcode number [2005ApPhA..80..753S]

30. Kabashin AV, Meunier M, Kingston C, Luong JHT: J. Phys. Chem. B. 2003, **107:** 4527. COI number [1:CAS:528:DC%2BD3sXjtlWhsrw%3D] COI number [1:CAS:528:DC%2BD3sXjtlWhsrw%3D]

31. Sylvestre J-P, Kabashin AV, Sacher E, Meunier M, Luong JHT: J. Am. Chem. Soc. (Commun.). 2004, **126:** 7176. COI number [1:CAS:528:DC%2BD2cXktVSlsL4%3D] COI number [1:CAS:528:DC%2BD2cXktVSlsL4%3D]

32. Kabashin AV, Meunier M: J. Photochem. Photobiol. A. 2006, **182:** 330–334. COI number [1:CAS:528:DC%2BD28Xot1Glsb8%3D] COI number [1:CAS:528:DC%2BD28Xot1Glsb8%3D]

33. Besner S, Kabashin AV, Winnik FW, Meunier M: Appl. Phys. A. 2008, **93:** 955–959. COI number [1:CAS:528:DC%2BD1cXht1Knsb7P]; Bibcode number [2008ApPhA..93..955B] COI number [1:CAS:528:DC%2BD1cXht1Knsb7P]; Bibcode number [2008ApPhA..93..955B]

34. Tokarev V, Marine W, Sentis M, Prat C: J. Appl. Phys.. 1995, **77:** 4714. COI number [1:CAS:528:DyaK2MXlt1ClsLg%3D]; Bibcode number [1995JAP....77.4714T] COI number [1:CAS:528:DyaK2MXlt1ClsLg%3D]; Bibcode number [1995JAP....77.4714T]

35. Chichkov BN, Momma C, Nolte S, Von Alvensleben F, Tünnermann A: Appl. Phys. A. 1996, **63:** 109. Bibcode number [1996ApPhA..63..109C] Bibcode number [1996ApPhA..63..109C]

36. Parisse JD, Marine M, Sentis J: Phys. IV 9. 1999, **9:** PR5–149.

37. Wu C, Crouch CH, Zhao L, Mazur E: Appl. Phys. Lett.. 2002, **81:** 1999. COI number [1:CAS:528:DC%2BD38XmvVSht78%3D]; Bibcode number [2002ApPhL..81.1999W] COI number [1:CAS:528:DC%2BD38XmvVSht78%3D]; Bibcode number [2002ApPhL..81.1999W]

38. Reif J, Costache F, Henyk M, Pandelov SV: Appl. Surf. Sci.. 2002, **197–198:** 891.

39. Daminelli G, Krüger J, Kautek W: Thin Solid Films. 2004, **467:** 334. COI number [1:CAS:528:DC%2BD2cXnsVGlsbs%3D]; Bibcode number [2004TSF...467..334D] COI number [1:CAS:528:DC%2BD2cXnsVGlsbs%3D]; Bibcode number [2004TSF...467..334D]

40. Besner S, Degorce J-Y, Kabashin AV, Meunier M: Appl. Surf. Sci.. 2005, **247:** 163–168. COI number [1:CAS:528:DC%2BD2MXltVOgs7k%3D]; Bibcode number [2005ApSS..247..163B] COI number [1:CAS:528:DC%2BD2MXltVOgs7k%3D]; Bibcode number [2005ApSS..247..163B]

41. Vorobyev AY, Guo C: Appl. Phys. Lett.. 2008, **92:** 041914. Bibcode number [2008ApPhL..92d1914V] Bibcode number [2008ApPhL..92d1914V]

42. Halbwax M, Sarnet T, Delaporte Ph, Sentis M, Etienne H, Torregrosa F, Vervisch V, Perichaud I, Martinuzzi S: Thin Solid Films. 2008, **516:** 6791. COI number [1:CAS:528:DC%2BD1cXnvVKitb0%3D]; Bibcode number [2008TSF...516.6791H] COI number [1:CAS:528:DC%2BD1cXnvVKitb0%3D]; Bibcode number [2008TSF...516.6791H]

43. Pereira A, Delaporte Ph, Sentis M, Cros A, Marine W, Basillais A, Thomann AL, Leborgne C, Semmar N, Andreazza P, Sauvage T: Thin Solid Films. 2004, **453–454:** 16.

44. Kabashin AV, Meunier M: Appl. Phys. Lett.. 2003, **82:** 1619. COI number [1:CAS:528:DC%2BD3sXhvV2ltL4%3D]; Bibcode number [2003ApPhL..82.1619K] COI number [1:CAS:528:DC%2BD3sXhvV2ltL4%3D]; Bibcode number [2003ApPhL..82.1619K]

45. Yang D, Kabashin A, Pilon-Marien V-G, Sacher E, Meunier M: J. Appl. Phys.. 2004, **95:** 5722. COI number [1:CAS:528:DC%2BD2cXjvVSrtL0%3D]; Bibcode number [2004JAP....95.5722Y] COI number [1:CAS:528:DC%2BD2cXjvVSrtL0%3D]; Bibcode number [2004JAP....95.5722Y]

46. Kabashin AV, Meunier M: Mat Sci. Eng. B. 2003, **101:** 60–64.

47. Kabashin AV, Magny F, Meunier M: J. Appl. Phys.. 2007, **101:** 054311. Bibcode number [2007JAP...101e4311K] Bibcode number [2007JAP...101e4311K]

48. Kabashin AV, Trudeau A, Marine W, Meunier M: Appl. Phys. Lett.. 2007, **91:** 201101. Bibcode number [2007ApPhL..91t1101K] Bibcode number [2007ApPhL..91t1101K]

49. Kabashin AV, Trudeau A, Marine W, Meunier M: Appl. Phys. A. 2008, **91:** 621. COI number [1:CAS:528:DC%2BD1cXlslKktbg%3D]; Bibcode number [2008ApPhA..91..621K] COI number [1:CAS:528:DC%2BD1cXlslKktbg%3D]; Bibcode number [2008ApPhA..91..621K]

50. Bunkin FV, Konov VI, Prokhorov AM, Fedorov VB: JTP Lett.. 1969, **9:** 371. Bibcode number [1969JETPL...9..371B] Bibcode number [1969JETPL...9..371B]

51. Itina T, Hermann J, Delaporte Ph, Sentis M: Appl. Surf. Sci.. 2003, **128:** 27.

52. Raizer YP: Laser-induced Discharge Phenomena. Consultants Bureau, New York; 1977.

53. Drouet MG, Pepin H: Appl. Phys. Lett.. 1976, **28:** 426. COI number [1:CAS:528:DyaE28XhslWqs7g%3D]; Bibcode number [1976ApPhL..28..426D] COI number [1:CAS:528:DyaE28XhslWqs7g%3D]; Bibcode number [1976ApPhL..28..426D]

54. Korobkin VV, Serov RV: Pisma Zh. Eksp. Teor. Fiz.. 1966, **4:** 103. [see also J. Exp. Theor. Phys. Lett. 4, 70 (1966)] [see also J. Exp. Theor. Phys. Lett. 4, 70 (1966)]

55. Kabashin AV, Nikitin PI: Quantum Electron.. 1997, **27:** 536. Bibcode number [1997QuEle..27..536K] Bibcode number [1997QuEle..27..536K]

56. Kabashin AV, Nikitin PI, Marine W, Sentis M: Appl. Phys. Lett.. 1998, **73:** 25. COI number [1:CAS:528:DyaK1cXktlOhsbk%3D]; Bibcode number [1998ApPhL..73...25K] COI number [1:CAS:528:DyaK1cXktlOhsbk%3D]; Bibcode number [1998ApPhL..73...25K]

57. Pereira A, Grojo D, Chaker M, Delaporte Ph, Guay D, Sentis M: Small. 2008, **4:** 572. COI number [1:CAS:528:DC%2BD1cXmvFWltLk%3D] COI number [1:CAS:528:DC%2BD1cXmvFWltLk%3D]

58. Piglmayer K, Denk R, Bauerle D: Appl. Phys. Lett.. 2002, **80:** 4693–4695. COI number [1:CAS:528:DC%2BD38XksleltL8%3D]; Bibcode number [2002ApPhL..80.4693P] COI number [1:CAS:528:DC%2BD38XksleltL8%3D]; Bibcode number [2002ApPhL..80.4693P]

59. Huang SM, Hong MH, Luk'yanchuk BS, Zheng YW, Song WD, Lu YF, Chong TC: J. Appl. Phys.. 2002, **92:** 2495–2500. COI number [1:CAS:528:DC%2BD38Xmt1Giu74%3D]; Bibcode number [2002JAP....92.2495H] COI number [1:CAS:528:DC%2 BD38Xmt1Giu74%3D]; Bibcode number [2002JAP....92.2495H]

60. Huang SM, Sun Z, Luk'yanchuk BS, Hong MH, Shi LP: Appl. Phys. Lett.. 2005, **86:** 161911. Bibcode number [2005ApPhL..86p1911H] Bibcode number [2005ApPhL..86p1911H]

61. Lu Y, Chen SC: Nanotechnology. 2003, **14:** 505–508. COI number [1:CAS:528:DC%2BD3sXlslart7c%3D]; Bibcode number [2003Nanot..14..505L] COI number [1:CAS:528:DC%2BD3sXlslart7c%3D]; Bibcode number [2003Nanot..14..505L]

62. Anker JN, Hall WP, Lyandres O, Shah NC, Zhao J, Van Duyne RP: Nature Mater.. 2008, **7:** 442–453. COI number [1:CAS:528:DC%2BD1cXmsVejt7g%3D]; Bibcode number [2008NatMa...7..442A] COI number [1:CAS:528:DC%2BD1cXmsVejt7g%3D]; Bibcode number [2008NatMa...7..442A]

63. Kabashin AV, Evans P, Patskovsky S, Wurtz G, Hendren W, Dickson W, Pollard RJ, Podolsky V, Zayats AV: Nature Mater.. 2009, **8:** 867–871. COI number [1:CAS:528:DC%2BD1MXhtlSgs77O]; Bibcode number [2009NatMa...8..867K] COI number [1:CAS:528:DC%2BD1MXhtlSgs77O]; Bibcode number [2009NatMa...8..867K]

64. Abraham FF: Homogeneous Nucleation Theory: The Pretransition Theory of Vapor Condensation. Academic Press, New York; 1974.

65. Kashchiev D: Nucleation: Basic Theory with Applications. Butterworth-Heinemann, Oxford; 2000.

66. Kabashin AV, Charbonneau-Lefort M, Meunier M, Leonelli R: Appl. Surf. Sci.. 2000, **168:** 328. COI number [1:CAS:528:DC%2BD3cXoslCqtbY%3D]; Bibcode number [2000ApSS..168..328K] COI number [1:CAS:528:DC%2BD3cXoslCqtbY%3D]; Bibcode number [2000ApSS..168..328K]

67. Kabashin AV, Meunier M, Leonelli R: J. Vacuum Sci. Tech. B. 2001, **19:** 2217. COI number [1:CAS:528:DC%2BD

3MXptFWntLg%3D] COI number [1:CAS:528:DC% 2BD3MX ptFWntLg%3D]

68. Prokes SM: Appl. Phys. Lett.. 1993, **62:** 3244. COI number [1:CAS:528:DyaK2cXmsV2qsg%3D%3D]; Bibcode number [1993ApPhL..62.3244P] COI number [1:CAS:528: DyaK2cXmsV2qsg%3D%3D]; Bibcode number [1993ApPhL. .62.3244P]

69. Kanemitsu Y, Ogawa T, Shiraishi K, Takeda K: Phys. Rev. B. 1993, **48:** 4883. COI number [1:CAS:528:DyaK3sXmtlOjsLg%3D]; Bibcode number [1993PhRvB..48.4883K] COI number [1:CAS: 528 :DyaK3sXmtlOjsLg%3D]; Bibcode number [1993PhRvB ..48.4883K]

70. Vogel A, Noack J, Nahen K, Theisen D, Busch S, Parlitz U, Hammer DX, Noojin GD, Rockwell BA, Birngruber R: Appl. Phys. B. 1999, **68:** 271. COI number [1:CAS:528: DyaK1MXnsFOitg%3D%3D]; Bibcode number [1999ApPhB..68..271V] COI number [1:CAS:528: DyaK1MXns FOitg%3D%3D]; Bibcode number [1999ApPhB..68..271V]

71. Sylvestre J-P, Poulin S, Kabashin AV, Sacher E, Meunier M, Luong JHT: J Phys Chem B. 2004, **108:** 16864–16869. COI number [1:CAS:528:DC%2BD2cXotFSqsb0%3D] COI number [1:CAS:528 :DC%2BD2cXotFSqsb0%3D] 10.1021/jp047134+

72. Besner S, Kabashin AV, Meunier M, Winnik FM: J Phys. Chem C.. 2009, **113:** 9526–9531. COI number [1:CAS:528:DC% 2BD1MXls1eitLw%3D] COI number [1:CAS:528:DC% 2BD1MXls1eitLw%3D] 10.1021/jp809275v

73. Rioux D, Laferriere M, Douplik A, Shah D, Lilge L, Kabashin AV, Meunier M: J. Biomed. Optics. 2009, **14:** 021010. Bibcode number [2009JBO....14b1010R] Bibcode number [2009JBO... .14b1010R]

Chapter 7

Three-Photon Absorption in ZnO Film Using Ultra Short Pulse Laser

Raied K. Jamal, Mohammed T. Hussein, Abdulla M. Suhail

Department of Physics, College of Science, University of Baghdad, Baghdad, Iraq

ABSTRACT

The three-photon absorption (3PA) in nanostructure wide-band gap ZnO semiconductor material is observed under high intensity femtosecond Titanium-Sapphire laser of 800 nm wavelength excitation. The ZnO film was prepared by chemical spray pyrolysis technique with substrate temperature of 400°C. The optical properties concerning the absorption, transmission, reflection, Raman and the photoluminescence spectra are studied for the prepared film. The structure of the ZnO film was tested with the X-ray diffraction and it was found to be a polycrystalline with recognized peaks oriented in (002), and (102). The measured of three photon absorption coefficient was found to be about 0.0166 $cm^3/Gwatt^2$, which is about five times higher than the bulk value. The fully computerized z-scan system was used to measure the nonlinear coefficients from the Gaussian fit of the transmitted laser incident.

Keywords: Multiphoton Processes; ZnO Nanocrystalline; Nonlinear Optics

1. INTRODUCTION

The ZnO nanostructures have many applications in gas sensors, UV detectors and solar cells [1-3]. This material has some additional advantages compared to other large band-gap semiconductors; for example, its large exciton binding energy (about 60 meV) which is three times the binding energies of ZnSe and GaN [4]. This allows a stable exciton distribution and achieves efficient excitonic emission at room temperature. The optical and electrical properties of ZnO nanostructures are studied at different preparation techniques and at different substrate materials [5,6]. The pumping of ZnO crystalline nanofilm with near-infrared femtosecond radiation pulses, enhances the nonlinear interaction between the ultrahigh intensity applied optical field and the ZnO nanostructures [7,8]. The nonlinear interaction leads to the simultaneous absorption of two or more photons of subband gap energy. The absorption is through a virtual-states that assists the inter band transitions. This transition produces electron-hole pairs in the excited states and, subsequently, the band-edge emission via their radiative recombination [9]. The two photon absorption (2PA) in semiconductor nanocrystals (NCs) has been widely investigated [8,10], while the research effort on their three-photon absorption (3PA) is limited [11]. The three photon absorption was observed in ZnO and ZnS crystals when pump with lasers of ultra excitation irradiance which was more than 40 GW/cm^2 [12-15]. Three-photon absorption in nanostructure wide-Band gap semiconductor ZnO using femtosecond laser for thin film was studied by [16].

In this work the three photon absorption in ZnO nanostructure illuminated by intensity 110 GW/cm^2 Ti-Sapphire laser is observed. The work concentrates on the effect of the nanostructure on the nonlinear parameters through the studying of the 3PA coefficient and the laser threshold pumping power. Simple mathematical relations are developed to describe the dependence of the fluorescent emission on the pumping laser intensity. The three photon absorption coefficient was calculated from the experimental measurements.

2. EXPERIMENTAL WORK

The ZnO film was prepared by chemical spray pyrolysis technique. The film was deposited on quartz substrates heated to 400°C. The spray solution is prepared by mixing Zinc Chloride ($ZnCl_2$) at 0.1 M with distillated water. The above mixture solution was placed in the flask of the atomizer and spread by controlled nitrogen gas flow on the heated substrates. The chemical spray pyrolysis experimental setup is similar to the standard unit fully described by [17]. The spraying time was controlled by adjustable a solenoid valve. The heated substrate was left for 12 sec after each spraying run to give time for the deposited ZnO layer to be dry. In order to get film of proper thickness many layers deposited of ZnO are required. The optimum experimental conditions for obtaining homogeneous ZnO film at 400°C are determined by the spraying time, the drying time and the flashing gas pressure. The schematic representation of the spray system is given in Figure 1. The possible reaction of the spray chemical on the heated substrate is yielding for the following reaction:

$$ZnCl_2 + H_2O \xrightarrow[400^{\circ}C]{Heat} ZnO\ (film/substrate) + 2HCl \qquad (1)$$

During the chemical reaction, gas and water vapor is obtained from this reaction due to the high temperature of the substrate. At end of reaction a white precipitates remain from the reaction as a nanofilm of ZnO as shown in Figure 2. The topography study of the prepared nanofilm surface was carried on using Scanning Electron Microscopy (SEM) type ULTRA 55 with different magnification; as shown in Figure 3. The X-ray diffraction (XRD) pattern of ZnO film was recorded by XDR 2000. The X-ray diffractometer use copper tube radiation line of wavelength 1.54Å in 2θ range from 20° to 60°. The scanning rate was 1 deg/min.

The UV-VIS-NIR absorption, transmission and reflection spectra of the sample were recorded by Hitachi U-4100 spectrometer covering the spectral range 200 - 1100 nm. The photoluminescence spectrum (PL) was studied using SL1174 spectrophotometer in the range 300 - 900 nm.

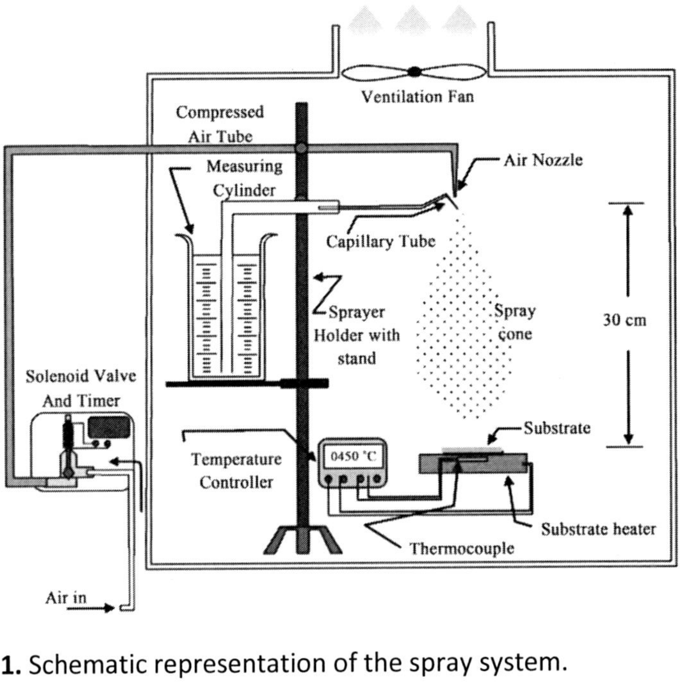

Figure 1. Schematic representation of the spray system.

Figure 2. A photo of spray pyrolyzed ZnO nanofilm on glass samples.

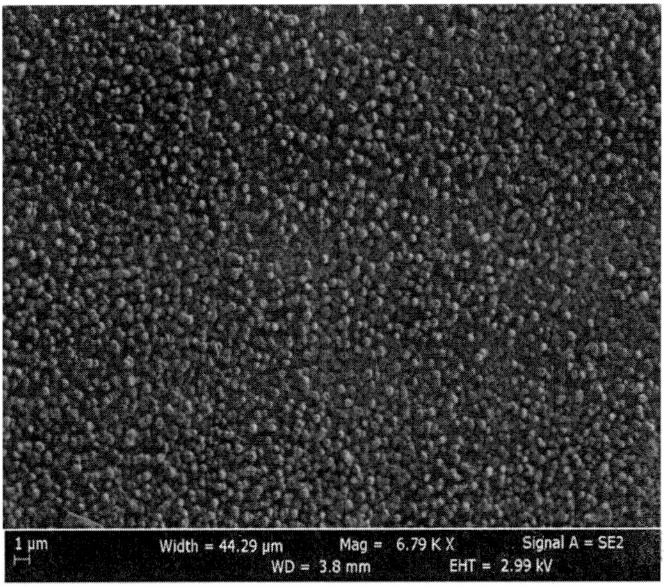
1 µm
Width = 44.29 µm
Mag = 6.79 K X
Signal A = SE2
WD = 3.8 mm
EHT = 2.99 kV

(a)

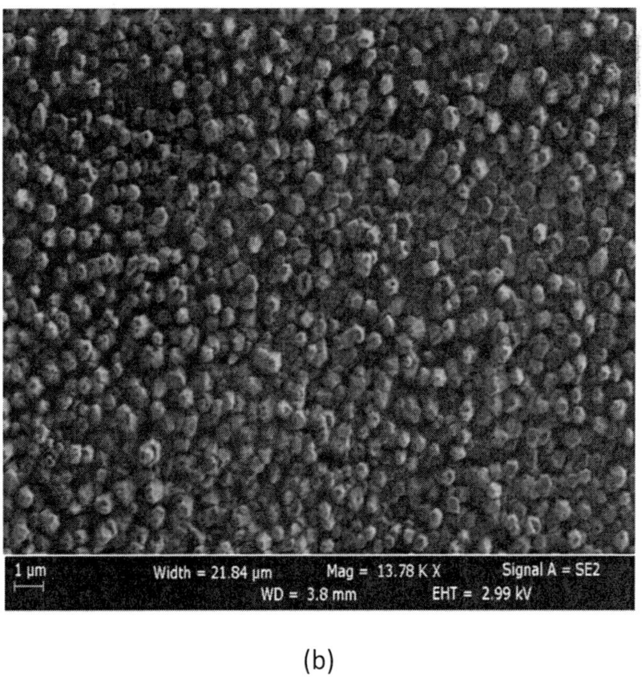
1 µm
Width = 21.84 µm
Mag = 13.78 K X
Signal A = SE2
WD = 3.8 mm
EHT = 2.99 kV

(b)

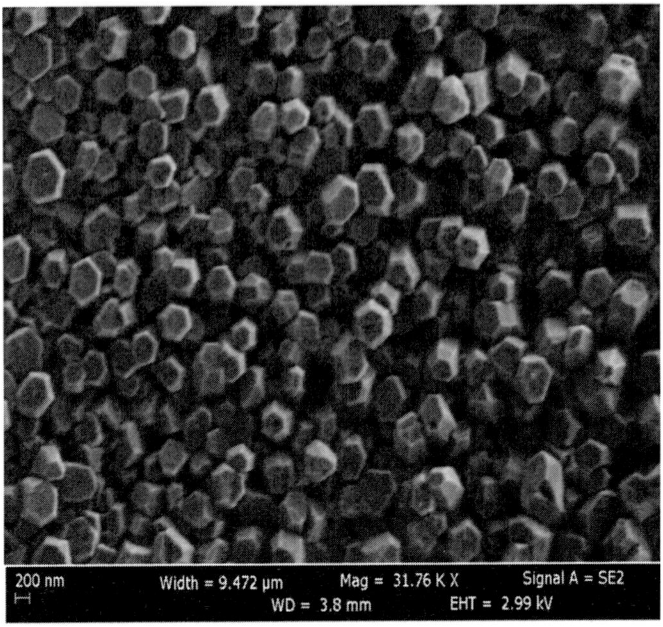
200 nm
Width = 9.472 µm
Mag = 31.76 K X
Signal A = SE2
WD = 3.8 mm
EHT = 2.99 kV

(c)

100 nm
Width = 4.011 µm
Mag = 75.00 K X
Signal A = SE2
WD = 6.8 mm
EHT = 2.99 kV

(d)

(e)

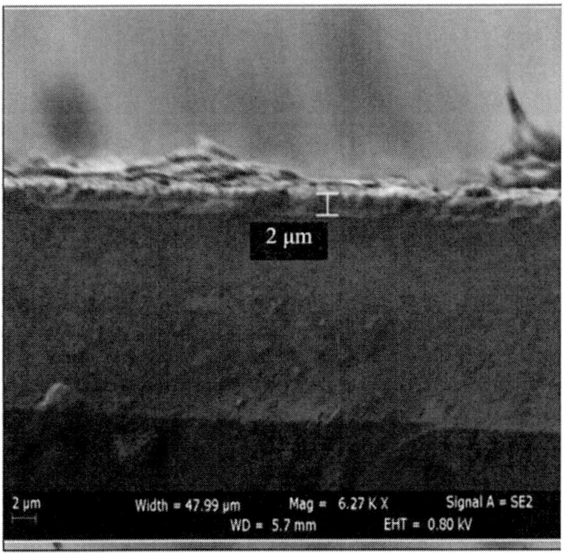

(f)

Figure 3. (a)-(e) represent SEM images of ZnO nanofilm at different magnification power and (f) for a sample thickness.

The nonlinear absorption study at the near resonant regime was carried out using single beam femtosecond z-scan technique. The z-scan setting is illustrated by schematic diagram shown in Figure 4. A femtosecond laser of pulse duration 60 fs and of average power 0.165 W was used as a laser source. The pulse duration was measured by autocorrelation system and the energy was measured by pyroelectric energy prob model type (PDA36A), covering the rang 350 - 1100 nm from THORLABS. The beam profile was adjusted by spatial filter leading to spatial intensity profile near Gaussian with beam quality of $M^2 \approx 1.79$. The laser beam was focused by a lens of 15 cm focal length to produce a waist of 22.5 μm. The sample was translated along the beam axis (z-axis) through the Rayleigh distance 2000 μm. The distance from lens to aperture was 53 cm and beam diameter at detector was 16.4 mm.

3. RESULTS AND DISCUSSION

The topography study of the prepared film shows the formation of the ZnO nanostructure and the film thickness was around 2 μm. as reveled by Scanning Electron Microscopy (SEM) type ULTRA 55 with different magnification; as shown in Figure 3. These figures showed nanocrystals of size ~(100 - 200 nm). The sample was scanned in all zones before the picture was taken. The micrographs reveled that the particles were hexagonal in shape. This indicated that ZnO nanocrystal were grown with c-axis orientation and almost perpendicular to the substrate surface as observed in the scanning electron microscope (SEM) image.

The XRD pattern of ZnO film prepared with 2 μm thickness is illustrated in Figure 3. The spectrum indicates that the ZnO film is a polycrystalline structure. The observed values of the XRD peaks are compared with American Society for testing and Material (ASTM) data for hexagonal zinc oxide. It can be noticed from the XRD pattern the strongest peak observed at Bragg's angle $2\theta = 34.38^\circ$ can be attributed to the (002) plan of the hexagonal ZnO and the interplanar distance can be determined by Bragg's law, $d_{002} = 0.26$ nm. The (102) peak is also observed at $2\theta = 47.44^\circ$ but this peak is much lower intensity than the

(002) peak. The figure shows broad peaks which give evidence of the nanostructure formation. Using the width of (002) peak which appears at angle 34.38° on 2θ scale in Scherrer's formula [18]:

$$d = 0.94\lambda/\beta\cos\theta \tag{2}$$

where d is the average crystalline grain size, λ is the wavelength, β represents the full width at half maximum (FWHM) in radian and θ is the Bragg diffraction angle in degree. The size of the formed nanoparticles was found to be about 46 nm, this value is close from the value that measured by SEM. Using the width of (102) peak which appears at angle 47.44° on 2θ scale the size of the formed nanoparticles was found to be about 26 nm. The interplaner distance can be determined by Bragg's law d_{102} = 0.19 nm.

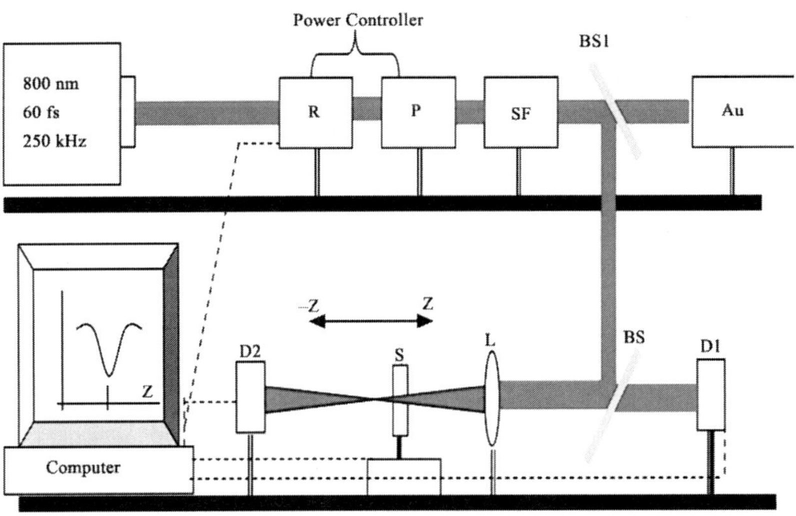

Figure 4. Schematic of the z-scan setup recording the nonlinear absorption, R-rotator, P-polarizer, SF-spatial filter, BS1, BS2-beam splitter, Au-autocorrelation, D1,D2-detectors, L-lens, S-sample.

The optical properties of ZnO film which were prepared on quartz substrate have been studied in this work. The absorption spectrum of the ZnO film is shown in Figure 5. This spectrum shows low absorption in the visible and infrared regions; however, the absorption in the ultraviolet region is high. This result were in agreement with the absorption spectrum which obtained by other workers. The energy band gap of ZnO film was estimated using Tauc equation which can be written as:

$$(\alpha h v) = A(hv - Eg)^n \tag{3}$$

where A is a constant, α absorption coefficient, hv the photon energy (Eg) the band gap, n = $1/2$ for the direct transitions. Referring to the data extracted from the absorption spectrum in Figure 6, the absorption coefficient (α) was calculated as a function of wavelength. Assuming allowed transition; the dependence of $(\alpha hv)^2$ on (hv) is plotted as in Figure 7. The extrapolation of the linear part of the plot $(\alpha hv)^2 = 0$, gives rise on estimation of the energy gap value of the prepared ZnO film. The value of the energy gap was found to be about 3.3 eV. This value was in a good agreement with values mentioned by other works.

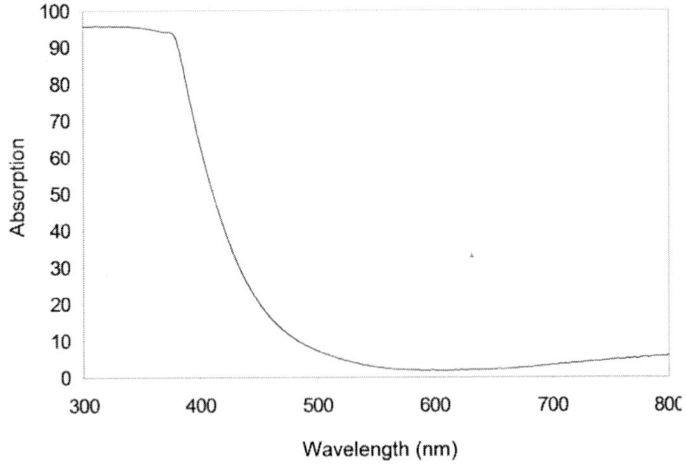

Figure 5. UV-VIS absorption spectra of ZnO nanofilm.

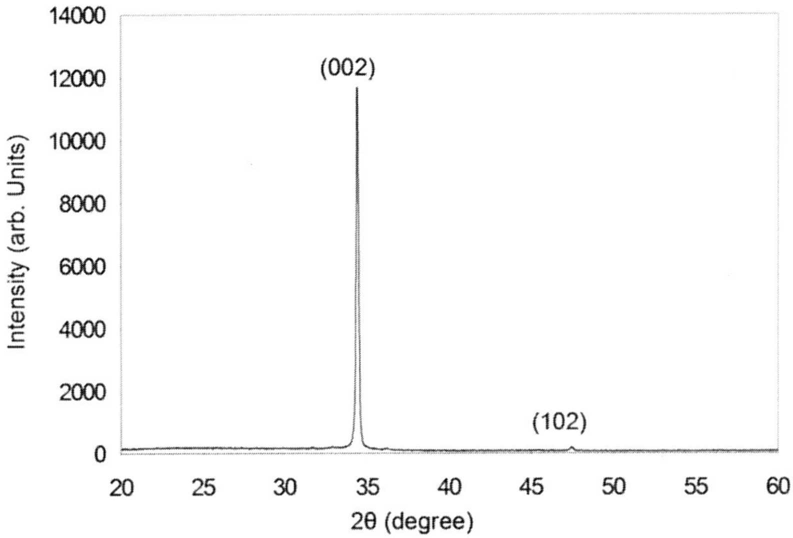

Figure 6. XRD pattern of ZnO nanofilm deposited on a glass substrate at 400°C.

The optical transmittance spectrum of the ZnO film is shown in Figure 8. It can be noticed from this figure that the transmittance is high in the visible and infrared regions and low transmission in ultraviolet region.

The optical reflection spectrum of the ZnO film is shown in Figure 9. It can be noticed from this figure that the maximum reflection at 470 nm and minimum value was 390 nm. The optical fluorescence spectrum of the ZnO film illuminated by 320 nm UV line is shown in Figure 10. The spectrum displays two major luminance peaks at 380 nm and around 556 nm. The first peak is due to the energy gap transmission which corresponds to 3.28 eV. The second peak is due to the excitonic emission; and it is in a good agreement with the results measured by many other authors [19,20]. The broad green emission peak that is dominated at 556 nm (≈2.23 eV) is a good evidence for the exciton formation in ZnO with a binding energy of 60 meV. The high binding energy enables the finding of the exciton at room temperature. The intensity at the 556 nm peak is higher than that found around 380 nm peak. This is because the band-to-band transition was quenched by the defect states. The same behavior was observed by [21].

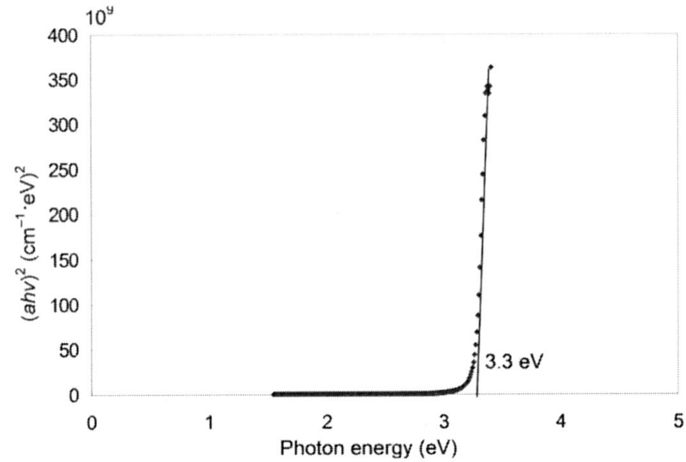

Figure 7. Plot of $(ahv)^2$ versus photon energy hv for ZnO nanonanofilm.

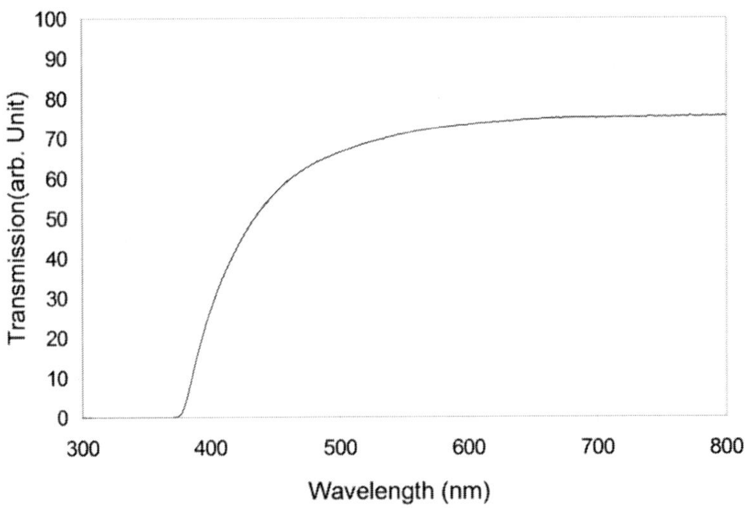

Figure 8. Optical transmission spectra of ZnO nanofilm.

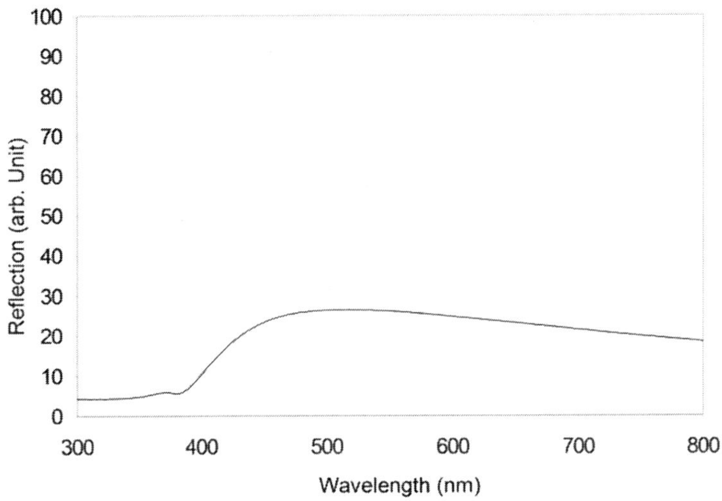

Figure 9. Optical reflection spectra of ZnO nanofilm.

The Raman active zone-center optical phonons predicated by the group theory are $A_1 + 2E_2 + E_1$. The phonons of A_1 and E_1 symmetry are polar phonons and, hence, exhibit different frequencies for the transverseoptical (TO) and longitudinal optical (LO) phonons. In wurtzite ZnO crystals, The non-polar phonon modes with symmetry E_2 have two frequencies, E_2(high) is associated with oxygen atoms and E_2(low) is associated with Zn sublattice. The described phonon modes have been reported in the Raman scattering spectra of bulk ZnO [22,23]. In Figure 11, we present a typical non-resonant Raman scattering spectrum from ZnO nanocrystal obtain under the 633 nm non-resonant excitation on of He-Ne laser. For comparison, the phonons peaks observed in bulk ZnO crystal are summarized in Table 1.

In our experiment, Raman spectra were recorded at room temperature under the 633 nm excitation and laser power 75 mW, the E_2(low) peak is red shifted by ~3 cm^{-1} from its bulk value, E_2(high) peak is red shift by ~2 cm^{-1}, but the peak E_1(LO) at 591 cm^{-1} corresponding to E_1(LO). The A_1(TO), E_1(TO), A_1(LO) peaks being absent in this sample.

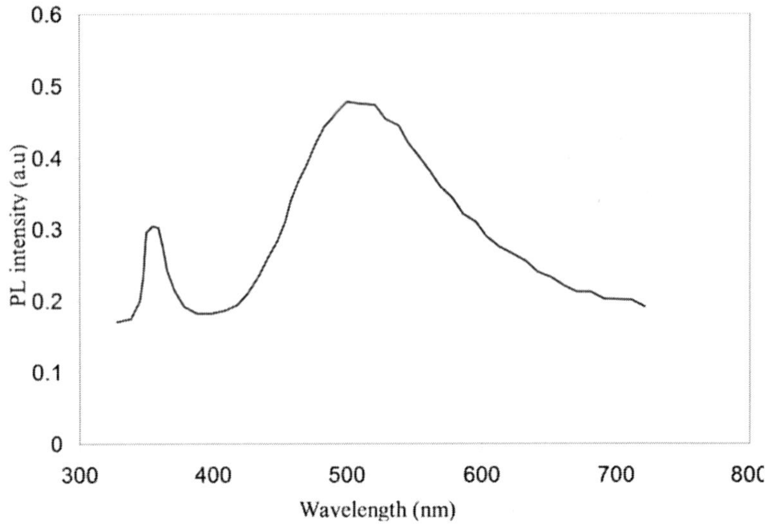

Figure 10. Photoluminescence emission spectrum of ZnO nanofilm.

There are three possible mechanisms for the phonon peak shift in Raman spectra of nanostructures. The first one is spatial confinement within the quantum dot (nanocrystal) boundaries. The second one is related to the phonon localization by defect. Nanocrystal or quantum dots, produced by chemical methods or by the molecularbeam epitaxy, normally have more defect than corresponding bulk crystal. In order elucidate the possible of the peak shifts we carried out simple calculations in the framework of the H. Richer [24] and I. H. Chambell [25] phenomenological models.

The z-scan transition curve at 110 GW/cm^2 excitation intensity is recorded for the nanoparticles ZnO film and it shown in Figure 12, Where the average power was 0.165 W, Repetition rate (RR) was 250 kHz, Pulse energy E_p was 6.6×10^{-7} J, Pulse duration (FWHM) was 60 fs, Peak power (PP) was 10.34×10^6 W.

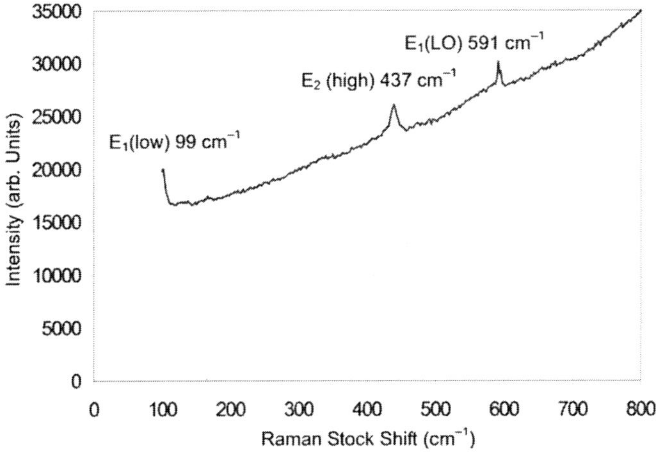

Figure 11. Non-resonant Raman spectra from ZnO nanocrystal under 633 nm excitation and 75 mW for He-Ne laser.

Table 1. Raman active phonon mode frequencies (in cm^{-1}) for bulk ZnO.

E_2(low)	A_1(TO)	E_1(TO)	E_2(high)	A_1(LO)	E_1(LO)
102	379	410	439	574	591

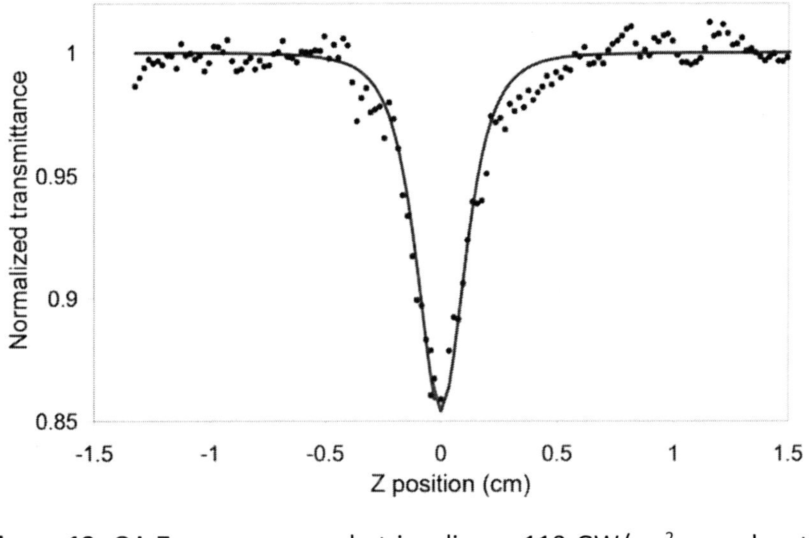

Figure 12. OA Z-scan measured at irradiance 110 GW/cm^2, wavelength 800 nm, a pulse duration 60 fs and repetition rate of 250 kHz. The solid line is the fitting curve employing the z-scan theory, described in the text, on 3PA.

The rate equations model describing the transitions between the valance band and the conduction band in our semiconductor sample illuminated by Titanium-Sapphire femtosecond laser of 800 nm wavelength can be written as:

$$\frac{dN_h}{dt} = -\frac{\alpha I^3}{3\hbar\omega} + \frac{N_e}{\tau_i} \tag{4}$$

$$\frac{dN_e}{dt} = \frac{\alpha I^3}{3\hbar\omega} - \frac{N_e}{\tau_i} \tag{5}$$

where N_h and N_e are the population in the valance band and the conduction band respectively. α is the three photon absorption coefficient, ħω is the energy of the IR pump Titanium-Sapphire laser photon, I is the laser intensity at the sample position and τ_i is the life time of the excited level in the conduction band. The Gaussian laser beam intensity is given by:

$$I = I_o \left[\frac{\omega_o^2}{\omega^2(z)}\right] \exp\left[-\frac{2r^2}{\omega^2}\right] \exp\left[-\frac{t^2}{\tau_p^2}\right] \tag{6}$$

where I_o is the laser intensity at the waist of the beam and τ_p is the laser pulse duration. Since the measuring intensity was carried out at the focal point, in the Z-coordinate, thus the intensity in Equation (4) can be reduced to $I = I_o$. In order to find the relation between the power of the out put emitted fluorescent pulse and the intensity of the Titanium-Sapphire laser pump, Equation (5) can be arranged as:

$$\frac{dN_e}{dt} + \frac{N_e}{\tau_i} = \frac{\alpha I_o^3}{3\hbar\omega} \tag{7}$$

The solution of Equation (7) can be written as:

$$N_e = \frac{\alpha \tau_i I_o^3}{3\hbar\omega}\left[1 - e^{-\frac{\tau_i}{\tau_p}}\right] \tag{8}$$

Since the pulse duration of the pump laser (τ_p) is in femtosecond and the life time of the excited state (τ_i) in picosecond [26], the Equation (8) can be simplified as:

$$N_e = \frac{\alpha \tau_p}{3\hbar\omega} I_o^3 \tag{9}$$

Multiply both sides of the above equation by the volume (V) of the illuminated part of the ZnO sample and arranged the terms, Equation (9) can be written as:

$$\frac{3\hbar\omega \cdot N_e \cdot V}{\tau_p} = P_f = \alpha \cdot V \cdot I_o^3 \tag{10}$$

where P_f is the power of the emitted fluorescent from the ZnO sample pumped by the Titanium-Sapphire laser. Introducing the quantum yield which include the probability of the exciton self-trapping and the efficiency of the detection unit, Equation (10) can be written as:

$$P_f = \eta \alpha V I_o^3 \tag{11}$$

In our experimental setting described before, the focal spot are about 1590 μm^2 and the ZnO film has 2 μm thickness, thus the illuminated volume is about 3180 μm^3 and η was estimated to by 10^{-5} in our experimental set up. The value of (a) is found from the measurements of P_f at different values of I_o, and by applying Equation (11), it was found to by about of 0.0142 cm^3/GWatt^2. The result which is not far away from the calculate result from Gaussian fit technique as follow:

The normalized energy transmittance for 3PA of the open aperture z-scan is given by R. L. Sutherland [27]:

$$T(z)=\frac{1}{\pi^{1/2}p_o} \times \int_{-\infty}^{\infty} \ln\left\{\left[1+p_o^2\exp\left(-2x^2\right)\right]^{1/2}+p_o\exp(-x^2)\right\}dx \tag{12}$$

where $P_o=\left(2\gamma I_o L'_{eff}\right)^{1/2}$, $I_o=I_{oo}/\left(1+z^2/z_o^2\right)$ is the excitation intensity at the position z, $z_o=\pi\omega_o^2/\lambda$ where z_o is the Rayleigh range, ω_o is the minimum beam waist at focal point (z = 0), λ is the laser free-space wavelength,

$L'_{eff}=\left[1-\exp(-2a_oL)\right]/2a_o$ is the effective sample length for 3PA processes; L is the sample length and a_o is the linear absorption coefficient. The open aperture z-scan graph is always normalized to linear transmittance i.e., transmittance at large values of |z|. If $P_o < 1$ Equation (12) can be expanded in a Taylor series as:

$$T=\sum_{m=1}^{\infty}(-1)^{m-1}\frac{p_o^{2m-2}}{(2m-1)!(2m-1)^{1/2}} \tag{13}$$

Furthermore, if the higher order terms are ignored, the transmission as a function of the incident intensity is given by R. L. Sutherland [27]:

$$T=1-\frac{\gamma I_o^2 L'_{eff}}{3^{3/2}} \tag{14}$$

The 3PA coefficient can be extracted from the best fit for Equation (14). The sold curve in Figure 12 is the best fit for Equation (14). The Equation (14) shows clearly that the depth of the absorption dip is linearly proportional to the 3PA coefficient γ, but the shape of the trace is primarily determined by the Rayleigh range of the focused Gaussian beam. The fitted value of γ is on the order of 0.0166 $cm^3/Gwatt^2$. This value is five times of magnitudes higher than the value observed with bulk ZnO sample. The natural logarithm of the (1 – T) values are plotted as a function of natural logarithm of the incident intensity I_o in Figure 13. the curve can be reasonably fitted with a straight line with slope of 1.99, this indicate that the 3PA was occur in ZnO pump by 800 nm laser source of 60 fs pulse duration.

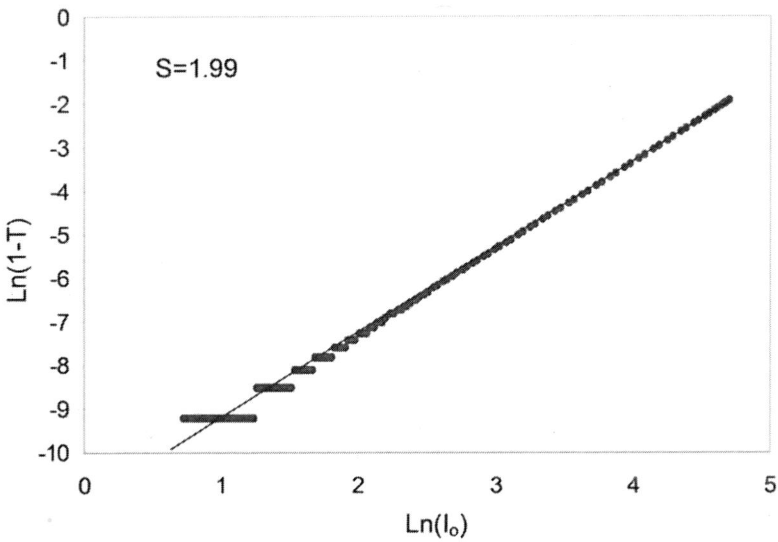

Figure 13. Plot of Ln(1 – T) versus Ln(I_o) at 800 nm wavelength, the solid line is the example of the linear at 800 nm with slope s = 1.99.

4. CONCLUSION

The three photon absorption has been observed in ZnO nanocrystalline prepared by chemical method upon illuminating it by femtosecond Titanium-Sapphire laser. The fully computerized z-scan system was used to measure the nonlinear absorption coefficient of the prepared samples. The value of the measured nonlinear coefficient was found to be five times higher than the bulk value. The enhancement of the nonlinear coefficient was attributed to the formation of the nanocrystallites of ZnO and to the existence of the exciton in the prepared film.

5. ACKNOWLEDGEMENTS

This work has been carried out in the physics Department, School of Engineering and Applied Sciences, Harvard University. The authors would like to thanks Mazur Research Group in Harvard University for their help through this work. Thank also to Christopher C. Evans, Jonthan D. B. Bradley, and Eric Mazur for their interest, guide and useful discussion. We thanks also the Ministry of Higher Education in the republic of Iraq for support this work.

REFERENCES

1. Y. w. Zhu, et al., "Multiwalled Carbon Nanotubes Beaded with ZnO nanoparticles for Ultrafast Nonlinear Optical Switching," Advanced Materials, Vol. 18, No. 5, 2006, pp. 587-592.

2. K. Ramanathan, J. Keane and R. Noufi, "Properties of High-Efficiency CIGS Thin-Film Solar Cells," NREL/CP, Vol. 520, 2005, Article ID: 37404.

3. C. L. Rhodes, S. Lappi, D. Fischer, S. Sambasivan, J. Genzer and S. Franzen, "Characterization of Monolayer Formation on Aluminum-Doped ZincOxide Thin Films," American Chemical Society, Vol. 24, No. 2, 2008, pp. 433-440.

4. H. Li, et al., "Microstructural Study of MBE-Grown ZnO Film on GaN/Sapphire (0001) Substrate," Central European Journal of Physics, Vol. 6, No. 3, 2008, pp. 638-642.

5. T. Sato, et al., "Production of Transition Metal-Doped ZnO Nanoparticles by Using RF Plasma Field," Journal of Crystal Growth, Vol. 275, No. 1-2, 2005, pp. 983-987.

6. S. Singh and M. S. R. Rao, "Structure and Physical Properties of Undoped ZnO and Vanadium Doped ZnO Films Deposited by Pulsed Laser Deposition," Journal Nanoscience and Nanotechnology, Vol. 8, No. 5, 2007, pp. 2575-2577.

7. Ü. Özgür, et al., "A Comprehensive Review of ZnO Materials and Devices," Journal of Applied Physics, Vol. 98, No. 4, 2005, Article ID: 041301.

8. J.-H. Lin, et al., "Two-Photon Resonance Assisted Huge Nonlinear Refraction and Absorption in ZnO Thin Films Institute of Electro-Optical Engineering," Journal of Applied Physics, Vol. 97, No. 3, 2005, Article ID: 033526.

9. S. J. Bentley, et al., "Three-Photon Absorption for Nanosecond Excitation in Cadmium Selenide Quantum Dots," Optical Engineering, Vol. 46, No. 12, 2007, Article ID: 128003. doi:10.1117/1.2823156

10. E. W. Van Stryiand, M. Sheik-Bahae, A. A. Said and D. J. Hagan, "Characterization of Nonlinear Optical Absorption and Refraction," Progress in Crystal Growth and Characterization of Materials, Vol. 27, No. 3-4, 1993, pp. 279-311.

11. J. He, W. Ji and J. Mi, "Three-Photon Absorption in Water-Soluble ZnS Nanocrystal," Applied Physics Letters, Vol. 88, No. 18, 2006, Article ID: 181114.

12. B. Gu, et al., "Three-Photon Absorption Saturation in ZnO and ZnS Crystals," Journal of Applied Physics, Vol. 103, No. 7, 2008, Article ID: 073105.

13. S. Pearl, et al., "Three Photon Absorption in Silicon for 2300 - 3300 nm," Applied Physics Letters, Vol. 93, No. 13, 2008, Article ID: 131102.

14. A. Penzkofer and W. Falkenstein, "Three Photon Absorption and Subsequent Excited-State Absorption in CdS," Optics Communications, Vol. 16, 1976, pp. 247-250.

15. M. G. Vivas, T. Shih, T. Voss, E. Mazur and C. R. Mendonca, "Nonlinear Spectra of ZnO: Reverse Saturable, Twoand Three-

Photon Absorption," Optics Express, Vol. 18, No. 9, 2010, pp. 9628-9633.

16. A M. Suhail, H. J. Kbashi and R. K. Jamal, "Three-Photon Absorption in Nanostructure Wide-Band Gap Semiconductor ZnO Using Femtosecond Laser," Modern Applied Science, Vol. 5, No. 6, 2011, pp. 199-210.

17. M. R. Islam and J. Podde, "Optical Properties of ZnO Nanofiber Thin Film Grown by Spray Pyrolysis of Zinc Acetate Precursor," Crystal Research and Technology, Vol. 44, No. 3, 2009, pp. 286-292.

18. A. L. Patterson, "The Scherrer Formula for X-Ray Particle Size Determination," Physical Reviews, Vol. 56, No. 10, 1939, pp. 978-982.

19. K. Yim and C. Lee, "Optical Properties of Al-Doped ZnO Thin Films Deposited by Two Different Sputtering Methods," Crystal Research and Technology, Vol. 41, No. 12, 2006, pp. 1198-1202.

20. C. X. Xu, et al., "Growth and Spectral Analysis of ZnO Nanotubes," Journal of Applied Physics, Vol. 103, No. 9, 2006, Article ID: 094303.

21. D. D. Wang, J. H. Yang, L. L. Yang, Y. J. Zhang, J. H. Lang and M. Gao, "Morphology and Photoluminescence Properties of ZnO Nanostructures Fabricated with Different Given Time of Ar," Crystal Research and Technology, Vol. 43, No. 10, 2008, pp. 1041-1045.

22. K. Alim, V. A. Fonoberov, M. Shamsa and A. A. Balandin, "Micro-Raman Investigation of Optical Phonons in ZnO Quantum Dots," Journal of Applied Physics, Vol. 97, No. 12, 2005, Article ID: 124313.

23. N. Ashkenov, B. N. Mbenkum, C. Bundesmann, V. Riede, M. Lorenz, D. Spenmann and E. M. Kaidashev, "Infrared Dielectric

Functions and Phonon Modes of High-Quality ZnO Film," Journal of Applied Physics, Vol. 93, No. 1, 2003, pp. 126-133.

24. H. Richer, Z. P. Wany, "The One Phone Raman Spectrum in Microcrystalline Silicon," Solid State Communications, Vol. 39, No. 5, 1981, pp. 625-629.

25. I. H. Chambell and p. M. Fanchet, "Three Effects of Microcrystal Size and Shape on the One Phonon Raman Spectra of Crystalline Semiconductors," Solid State Communications, Vol. 58, 1986, p. 739.

26. J. He, et al., "Three-Photon Absorption in ZnO and ZnS Crystals," Optics Express, Vol. 13, 2005, pp. 9235-9247.

27. R. L. Sutherland, D. G. McLean and S. Kirkpatrick, "Handbook of Nonlinear Optics," 2nd Edition, Reserved and Expanded, New York, 2003.

Chapter 8

Calculation of Atomic Data and Gain Coefficient for XUV & Soft X-Ray Laser Emission from Ge XXIII

Wessameldin S. Abdelaziz

National Institute of Laser Enhanced Sciences, Cairo University, Giza, Egypt

ABSTRACT

Energy levels, transition probabilities, oscillator strengths, and collision strengths have been calculated for transitions in Ne-like Ge. The data refer to a 241 fine-structure levels belonging to the configurations 1s2 2s2 2p5nl, 1s2 2s1 2p6nl (n = 3, 4, 5, 6; l = s, p, d, f, g and h), which have been calculated by the fully relativistic flexible atomic code (FAC). These data are used in the determination of the reduced population for the 241 fine structure levels and gain coefficients over a wide range of electron densities (from 2 × 10+20 to 4 × 10+22 cm−3) and at various electron plasma temperatures (650, 850, 1050, 1250, 1450, 1650, 1850) eV by using the MATLAB R2013a Computer program for solving simultaneous coupled rate equations. The reduced population for the 241 fine structure levels the gain coefficients for those transitions with positive population inversion factor are determined and plotted against the electron density.

Keywords: Energy Levels, Transition Probabilities, Oscillator Strengths, Collision Strengths

1. INTRODUCTION

The concept for x-ray lasers went back to the 1965s, when it was based on the laser-produced plasmas by electron de-excitation having a promising interest science, which was first proposed by the Soviet scientists Gudzenko and Shelepin [1] . They thought that the short wavelength laser in the x-ray region of the electromagnetic spectrum needed a large energy gap which was sustained in the highly ionized ions. After ionization, in the equilibrium plasma, ions having specific number of electrons such as 2 (helium-like), 10 (neon-like), 28 (nickel-like) and 46 (palladium-like) were relatively stable and survived in a wide range of temperature and density [2] [3] . X-ray lasers pumping methods are electron collisional excitation, photo excitation, charge transfer, electron collisional recombination and de-electronic recombination pumping using a picosecond chirped pulse amplification (CPA) pulse [4] , a capillary discharge [5] [6] and a free electron laser. However, the electron collisional excitation pumping of the inner shell and outer shell of highly ionization states have shown a more stable and a higher output [7] [8] . Thus, the electron collisional pumping proved to be the most efficient method in producing the x-ray lasers.

Experimentally there exist in the literature some studies trying to develop high-efficiency X-ray laser with significant gain, for example, [3] [9] proposed the original mechanism for demonstrating X-ray lasing by resonant photo pumping. Several authors during the past three decades [10] [11] have studied this lasing mechanism experimentally and theoretically, in the hope of developing high-efficiency X-ray laser. In another study by Qi, N. and Krishnan, M. [12] , the shortest wavelength at which the significant gain had been measured using the resonant photo pumping was in the beryllium-like carbon at 2163 Å, which was far from the X-ray spectral region.

Table 1. Energy levels of Ge XXIII, and their threshold energies (in eV).

Index	State Configuration	J^{π}	FAC	COWAN	NIST	Exp.	CIV3	MCDF	RCI
1	$2p^5$	0^e	0.00000E+00	0.00000E+00	0.00000E+00	0.00000E+00	0.00000E+00	0.00000E+00	0.00000E+00
2	$2p_{3/2}3s_{1/2}$	2°	1.23421E+03	1.23585E+03		1.236122E+03	1.23596E+03	1.234804E+03	1.236931E+03
3	$2p_{3/2}3s_{1/2}$	1°	1.23709E+03	1.23863E+03	1.23900E+03	1.238602E+03	1.23859E+03	1.237659E+03	1.239498E+03
4	$2p_{1/2}3s_{1/2}$	0°	1.26664E+03	1.26835E+03		1.268358E+03	1.26828E+03	1.267259E+03	1.269494E+03
5	$2p_{1/2}3s_{1/2}$	1°	1.26822E+03	1.26994E+03	1.27009E+03	1.269598E+03	1.26978E+03	1.268864E+03	1.270625E+03
6	$2p_{3/2}3p_{1/2}$	1^e	1.27690E+03	1.27848E+03	---	1.278277E+03	1.27857E+03	1.277620E+03	1.280475E+03
7	$2p_{3/2}3p_{1/2}$	2^e	1.28046E+03	1.28213E+03	---	1.281997E+03	1.28213E+03	1.280943E+03	1.283104E+03
8	$2p_{3/2}3p_{3/2}$	3^e	1.28642E+03	1.28801E+03	---	1.288196E+03	1.28801E+03	1.286986E+03	1.288182E+03
9	$2p_{3/2}3p_{3/2}$	1^e	1.28726E+03	1.28859E+03	---	1.288196E+03	1.28869E+03	1.287800E+03	1.288267E+03
10	$2p_{3/2}3p_{3/2}$	2^e	1.29059E+03	1.29201E+03	1.29204E+03	1.291915E+03	1.29190E+03	1.291035E+03	1.291087E+03
11	$2p_{3/2}3p_{3/2}$	0^e	1.30216E+03	1.30328E+03	---	1.303074E+03	1.30354E+03	1.302522E+03	1.301413E+03
12	$2p_{1/2}3p_{1/2}$	1^e	1.31186E+03	1.31352E+03	---	1.314232E+03	1.31355E+03	1.312405E+03	1.314407E+03
13	$2p_{1/2}3p_{3/2}$	1^e	1.31985E+03	1.32137E+03	---	1.321671E+03	1.32122E+03	1.320429E+03	1.322501E+03
14	$2p_{1/2}3p_{3/2}$	2^e	1.32083E+03	1.32231E+03	1.32254E+03	1.322911E+03	1.32229E+03	1.321340E+03	1.323593E+03
15	$2p_{1/2}3p_{1/2}$	0^e	1.33447E+03	1.33333E+03	1.33333E+03	1.332830E+03	1.33321E+03	1.332304E+03	1.334537E+03
16	$2p_{3/2}3d_{3/2}$	0°	1.34070E+03	1.34234E+03	---	1.342749E+03	1.34218E+03	1.341534E+03	1.343407E+03
17	$2p_{3/2}3d_{3/2}$	1°	1.34247E+03	1.34422E+03	1.34473E+03	1.343989E+03	1.34403E+03	1.343296E+03	1.345574E+03
18	$2p_{3/2}3d_{5/2}$	4°	1.34553E+03	1.34748E+03	---	---	1.34724E+03	1.346399E+03	1.348306E+03
19	$2p_{3/2}3d_{5/2}$	2°	1.34575E+03	1.34753E+03	---	1.347708E+03	1.34730E+03	1.346560E+03	1.348784E+03
20	$2p_{3/2}3d_{3/2}$	3°	1.34594E+03	1.34772E+03	---	1.347708E+03	1.34779E+03	1.346843E+03	1.348794E+03
21	$2p_{3/2}3d_{3/2}$	2°	1.34871E+03	1.35031E+03	---	1.350188E+03	1.35025E+03	1.349561E+03	1.351217E+03

Continued

22	$2p_{3/2}3d_{5/2}$	3°	1.35077E+03	1.35267E+03	---	1.352668E+03	1.35230E+03	1.351588E+03	1.353658E+03
23	$2p_{3/2}3d_{5/2}$	1°	1.35938E+03	1.35991E+03	1.36085E+03	1.361346E+03	1.36075E+03	1.360284E+03	1.362151E+03
24	$2p_{1/2}3d_{3/2}$	2°	1.37811E+03	1.37920E+03	---	---	1.37955E+03	1.378922E+03	1.380718E+03
25	$2p_{1/2}3d_{5/2}$	2°	1.37971E+03	1.38104E+03	---	---	1.38101E+03	1.380575E+03	1.382465E+03
26	$2p_{1/2}3d_{5/2}$	3°	1.38086E+03	1.38195E+03	---	1.383664E+03	1.38231E+03	1.381764E+03	1.383652E+03
27	$2p_{1/2}3d_{3/2}$	1°	1.38931E+03	1.38884E+03	1.38825E+03	1.391103E+03	1.39058E+03	1.389940E+03	1.392453E+03
28	$2s_{1/2}3s_{1/2}$	1^e	1.42808E+03	---	---	---	1.40081E+03	---	1.430234E+03
29	$2s_{1/2}3s_{1/2}$	0^e	1.43744E+03	---	---	---	1.40856E+03	---	1.438962E+03
30	$2s_{1/2}3p_{1/2}$	0°	1.47300E+03	---	---	---	1.43866E+03	---	1.474996E+03
31	$2s_{1/2}3p_{1/2}$	1°	1.47381E+03	---	1.47132E+03	---	1.43962E+03	---	1.476085E+03
32	$2s_{1/2}3p_{3/2}$	2°	1.48066E+03	---	---	---	1.44646E+03	---	1.482977E+03
33	$2s_{1/2}3p_{3/2}$	1°	1.48318E+03	---	1.48074E+03	---	1.44867E+03	---	1.485577E+03
34	$2s_{1/2}3d_{3/2}$	1^e	1.53678E+03	---	---	---	1.49950E+03	---	1.539105E+03
35	$2s_{1/2}3d_{3/2}$	2^e	1.53722E+03	---	---	---	1.50003E+03	---	1.539522E+03
36	$2s_{1/2}3d_{5/2}$	3^e	1.53813E+03	---	---	---	1.50093E+03	---	1.540470E+03
37	$2s_{1/2}3d_{5/2}$	2^e	1.54524E+03	---	---	---	1.50787E+03	---	1.547695E+03
38	$2p_{3/2}4s_{1/2}$	2°	1.67001E+03	1.67259E+03	---	---	---	---	---
39	$2p_{3/2}4s_{1/2}$	1°	1.67092E+03	1.67259E+03	1.67250E+03	---	---	---	---
40	$2p_{3/2}4p_{1/2}$	1^e	1.68781E+03	1.69062E+03	---	---	---	---	---
41	$2p_{3/2}4p_{1/2}$	2^e	1.68868E+03	1.69125E+03	---	---	---	---	---
42	$2p_{3/2}4p_{3/2}$	3^e	1.69117E+03	1.69380E+03	---	---	---	---	---
43	$2p_{3/2}4p_{3/2}$	1^e	1.69150E+03	1.69405E+03	---	---	---	---	---
44	$2p_{3/2}4p_{3/2}$	2^e	1.69258E+03	1.69497E+03	---	---	---	---	---

45	$2p_{3/2}4p_{3/2}$	0^e	1.69841E+03	1.69999E+03	---	---	---	---	---
46	$2p_{1/2}4s_{1/2}$	0°	1.70254E+03	1.70497E+03	---	---	---	---	---
47	$2p_{1/2}4s_{1/2}$	1°	1.70295E+03	1.70497E+03	1.70466E+03	---	---	---	---
48	$2p_{3/2}4d_{3/2}$	0°	1.71168E+03	1.71460E+03	---	---	---	---	---
49	$2p_{3/2}4d_{3/2}$	1°	1.71237E+03	1.71521E+03	---	---	---	---	---
50	$2p_{3/2}4d_{5/2}$	4°	1.71336E+03	1.71608E+03	---	---	---	---	---
51	$2p_{3/2}4d_{3/2}$	3°	1.71340E+03	1.71621E+03	---	---	---	---	---
52	$2p_{3/2}4d_{5/2}$	2°	1.71356E+03	1.71631E+03	---	---	---	---	---
53	$2p_{3/2}4d_{3/2}$	2°	1.71440E+03	1.71704E+03	---	---	---	---	---
54	$2p_{3/2}4d_{5/2}$	3°	1.71516E+03	1.71772E+03	---	---	---	---	---
55	$2p_{3/2}4d_{5/2}$	1°	1.71893E+03	1.72055E+03	1.72078E+05	---	---	---	---
56	$2p_{1/2}4p_{1/2}$	1^e	1.72080E+03	1.72379E+03	---	---	---	---	---
57	$2p_{1/2}4p_{3/2}$	1^e	1.72399E+03	1.72689E+03	---	---	---	---	---
58	$2p_{1/2}4p_{3/2}$	2^e	1.72433E+03	1.72708E+03	---	---	---	---	---
59	$2p_{3/2}4f_{5/2}$	1^e	1.72480E+03	1.72713E+03	---	---	---	---	---
60	$2p_{3/2}4f_{5/2}$	4^e	1.72501E+03	1.72742E+03	---	---	---	---	---
61	$2p_{3/2}4f_{7/2}$	2^e	1.72515E+03	1.72746E+03	---	---	---	---	---

62	$2p_{3/2}4f_{7/2}$	5^e	1.72516E+03	1.72764E+03	---	---	---	---	---
63	$2p_{3/2}4f_{7/2}$	3^e	1.72574E+03	1.72798E+03	---	---	---	---	---
64	$2p_{3/2}4f_{5/2}$	2^e	1.72585E+03	1.72807E+03	---	---	---	---	---
65	$2p_{3/2}4f_{5/2}$	3^e	1.72591E+03	1.72810E+03	---	---	---	---	---
66	$2p_{3/2}4f_{7/2}$	4^e	1.72613E+03	1.72817E+03	---	---	---	---	---
67	$2p_{1/2}4p_{1/2}$	0^e	1.72618E+03	1.72838E+03	---	---	---	---	---
68	$2p_{1/2}4d_{3/2}$	2°	1.74573E+03	1.74855E+03	---	---	---	---	---
69	$2p_{1/2}4d_{5/2}$	2°	1.74642E+03	1.74941E+03	---	---	---	---	---
70	$2p_{1/2}4d_{5/2}$	3°	1.74687E+03	1.74986E+03	---	---	---	---	---
71	$2p_{1/2}4d_{3/2}$	1°	1.74908E+03	1.75109E+03	1.75090E+03	---	---	---	---
72	$2p_{1/2}4f_{5/2}$	3^e	1.75784E+03	1.76052E+03	---	---	---	---	---
73	$2p_{1/2}4f_{7/2}$	4^e	1.75816E+03	1.76073E+03	---	---	---	---	---
74	$2p_{1/2}4f_{5/2}$	2^e	1.75818E+03	1.76084E+03	---	---	---	---	---
75	$2p_{1/2}4f_{7/2}$	3^e	1.75824E+03	1.76085E+03	---	---	---	---	---
76	$2p_{3/2}5s_{1/2}$	2°	1.86014E+03	---	---	---	---	---	---
77	$2p_{3/2}5s_{1/2}$	1°	1.86061E+03	---	---	---	---	---	---
78	$2s_{1/2}4s_{1/2}$	1^e	1.86083E+03	---	---	---	---	---	---
79	$2s_{1/2}4s_{1/2}$	0^e	1.86447E+03	---	---	---	---	---	---
80	$2p_{3/2}5p_{1/2}$	2^e	1.86954E+03	---	---	---	---	---	---
81	$2p_{3/2}5p_{1/2}$	1^e	1.86965E+03	---	---	---	---	---	---
82	$2p_{3/2}5p_{3/2}$	3^e	1.87079E+03	---	---	---	---	---	---
83	$2p_{3/2}5p_{3/2}$	1^e	1.87117E+03	---	---	---	---	---	---

84	$2p_{3/2}5p_{3/2}$	2^e	1.87148E+03	---	---	---	---	---	---
85	$2p_{3/2}5p_{3/2}$	0^e	1.87450E+03	---	---	---	---	---	---
86	$2s_{1/2}4p_{1/2}$	0°	1.87921E+03	---	---	---	---	---	---
87	$2s_{1/2}4p_{1/2}$	1°	1.87983E+03	---	---	---	---	---	---
88	$2p_{3/2}5d_{3/2}$	0°	1.88142E+03	---	---	---	---	---	---
89	$2p_{3/2}5d_{3/2}$	1°	1.88158E+03	---	1.88679E+03	---	---	---	---
90	$2p_{3/2}5d_{5/2}$	2°	1.88163E+03	---	---	---	---	---	---
91	$2p_{3/2}5d_{3/2}$	3°	1.88172E+03	---	---	---	---	---	---
92	$2p_{3/2}5d_{5/2}$	4°	1.88173E+03	---	---	---	---	---	---
93	$2p_{3/2}5d_{3/2}$	2°	1.88217E+03	---	---	---	---	---	---
94	$2p_{3/2}5d_{5/2}$	3°	1.88255E+03	---	---	---	---	---	---
95	$2s_{1/2}4p_{3/2}$	2°	1.88335E+03	---	---	---	---	---	---
96	$2s_{1/2}4p_{3/2}$	1°	1.88338E+03	---	1.88679E+03	---	---	---	---
97	$2p_{3/2}5d_{5/2}$	1°	1.88476E+03	---	---	---	---	---	---
98	$2p_{3/2}5f_{5/2}$	1^e	1.88719E+03	---	---	---	---	---	---
99	$2p_{3/2}5f_{5/2}$	2^e	1.88742E+03	---	---	---	---	---	---
100	$2p_{3/2}5f_{5/2}$	4^e	1.88754E+03	---	---	---	---	---	---
101	$2p_{3/2}5f_{7/2}$	5^e	1.88760E+03	---	---	---	---	---	---

102	$2p_{3/2}5f_{7/2}$	3^e	1.88780E+03	---	---	---	---	---	---
103	$2p_{3/2}5g_{7/2}$	2°	1.88786E+03	---	---	---	---	---	---
104	$2p_{3/2}5g_{9/2}$	3°	1.88795E+03	---	---	---	---	---	---
105	$2p_{3/2}5f_{5/2}$	3^e	1.88795E+03	---	---	---	---	---	---
106	$2p_{3/2}5f_{7/2}$	2^e	1.88797E+03	---	---	---	---	---	---
107	$2p_{3/2}5f_{7/2}$	4^e	1.88808E+03	---	---	---	---	---	---
108	$2p_{3/2}5g_{7/2}$	5°	1.88817E+03	---	---	---	---	---	---
109	$2p_{3/2}5g_{7/2}$	3°	1.88824E+03	---	---	---	---	---	---
110	$2p_{3/2}5g_{9/2}$	6°	1.88825E+03	---	---	---	---	---	---
111	$2p_{3/2}5g_{9/2}$	4°	1.88830E+03	---	---	---	---	---	---
112	$2p_{3/2}5g_{7/2}$	4°	1.88842E+03	---	---	---	---	---	---
113	$2p_{3/2}5g_{9/2}$	5°	1.88849E+03	---	---	---	---	---	---
114	$2p_{1/2}5s_{1/2}$	0°	1.89308E+03	---	---	---	---	---	---
115	$2p_{1/2}5s_{1/2}$	1°	1.89329E+03	---	---	---	---	---	---
116	$2p_{1/2}5p_{1/2}$	1^e	1.90155E+03	---	---	---	---	---	---
117	$2p_{1/2}5p_{3/2}$	2^e	1.90316E+03	---	---	---	---	---	---
118	$2p_{1/2}5p_{3/2}$	1^e	1.90352E+03	---	---	---	---	---	---
119	$2p_{1/2}5p_{1/2}$	0^e	1.90422E+03	---	---	---	---	---	---
120	$2s_{1/2}4d_{5/2}$	3^e	1.90492E+03	---	---	---	---	---	---
121	$2s_{1/2}4d_{3/2}$	1^e	1.90493E+03	---	---	---	---	---	---
122	$2s_{1/2}4d_{3/2}$	2^e	1.90495E+03	---	---	---	---	---	---
123	$2s_{1/2}4d_{3/2}$	2^e	1.90740E+03	---	---	---	---	---	---
124	$2p_{1/2}5d_{3/2}$	2°	1.91418E+03	---	---	---	---	---	---

125	$2p_{1/2}5d_{5/2}$	2°	1.91454E+03	---	---	---	---	---	---
126	$2p_{1/2}5d_{5/2}$	3°	1.91471E+03	---	---	---	---	---	---
127	$2s_{1/2}4f_{5/2}$	3°	1.91553E+03	---	---	---	---	---	---
128	$2s_{1/2}4f_{7/2}$	4°	1.91561E+03	---	---	---	---	---	---
129	$2p_{1/2}5d_{3/2}$	1°	1.91567E+03	---	1.91779E+03	---	---	---	---
130	$2s_{1/2}4f_{5/2}$	2°	1.91624E+03	---	---	---	---	---	---
131	$2s_{1/2}4f_{7/2}$	3°	1.91667E+03	---	---	---	---	---	---
132	$2p_{1/2}5f_{5/2}$	3^{e}	1.92023E+03	---	---	---	---	---	---
133	$2p_{1/2}5f_{7/2}$	4^{e}	1.92040E+03	---	---	---	---	---	---
134	$2p_{1/2}5f_{5/2}$	2^{e}	1.92046E+03	---	---	---	---	---	---
135	$2p_{1/2}5f_{7/2}$	3^{e}	1.92049E+03	---	---	---	---	---	---
136	$2p_{1/2}5g_{7/2}$	4°	1.92087E+03	---	---	---	---	---	---
137	$2p_{1/2}5g_{9/2}$	5°	1.92095E+03	---	---	---	---	---	---
138	$2p_{1/2}5g_{7/2}$	3°	1.92159E+03	---	---	---	---	---	---
139	$2p_{1/2}5g_{9/2}$	4°	1.92168E+03	---	---	---	---	---	---
140	$2p_{3/2}6s_{1/2}$	2°	1.96032E+03	---	---	---	---	---	---
141	$2p_{3/2}6s_{1/2}$	1°	1.96055E+03	---	---	---	---	---	---

142	$2p_{3/2}6p_{1/2}$	1^e	1.96543E+03	---	---	---	---	---	---
143	$2p_{3/2}6p_{1/2}$	2^e	1.96562E+03	---	---	---	---	---	---
144	$2p_{3/2}6p_{3/2}$	3^e	1.96636E+03	---	---	---	---	---	---
145	$2p_{3/2}6p_{3/2}$	1^e	1.96644E+03	---	---	---	---	---	---
146	$2p_{3/2}6p_{3/2}$	2^e	1.96670E+03	---	---	---	---	---	---
147	$2p_{3/2}6p_{3/2}$	0^e	1.96836E+03	---	---	---	---	---	---
148	$2p_{3/2}6d_{3/2}$	0°	1.97209E+03	---	---	---	---	---	---
149	$2p_{3/2}6d_{3/2}$	1°	1.97229E+03	---	---	---	---	---	---
150	$2p_{3/2}6d_{3/2}$	3°	1.97253E+03	---	---	---	---	---	---
151	$2p_{3/2}6d_{5/2}$	4°	1.97257E+03	---	---	---	---	---	---
152	$2p_{3/2}6d_{3/2}$	2°	1.97259E+03	---	---	---	---	---	---
153	$2p_{3/2}6d_{5/2}$	2°	1.97279E+03	---	---	---	---	---	---
154	$2p_{3/2}6d_{5/2}$	3°	1.97299E+03	---	---	---	---	---	---
155	$2p_{3/2}6d_{5/2}$	1°	1.97404E+03	---	1.97680E+03	---	---	---	---
156	$2p_{3/2}6f_{5/2}$	1^e	1.97568E+03	---	---	---	---	---	---
157	$2p_{3/2}6f_{5/2}$	4^e	1.97579E+03	---	---	---	---	---	---
158	$2p_{3/2}6f_{5/2}$	2^e	1.97580E+03	---	---	---	---	---	---
159	$2p_{3/2}6f_{7/2}$	5^e	1.97583E+03	---	---	---	---	---	---
160	$2p_{3/2}6f_{5/2}$	3^e	1.97597E+03	---	---	---	---	---	---
161	$2p_{3/2}6f_{7/2}$	3^e	1.97602E+03	---	---	---	---	---	---
162	$2p_{3/2}6p_{7/2}$	2^e	1.97605E+03	---	---	---	---	---	---
163	$2p_{3/2}6f_{7/2}$	4^e	1.97609E+03	---	---	---	---	---	---
164	$2p_{3/2}6g_{7/2}$	2°	1.97626E+03	---	---	---	---	---	---

165	$2p_{3/2}6g_{7/2}$	5°	1.97626E+03	---	---	---	---	---	---
166	$2p_{3/2}6h_{9/2}$	3^{e}	1.97627E+03	---	---	---	---	---	---
167	$2p_{3/2}6h_{11/2}$	4^{e}	1.97630E+03	---	---	---	---	---	---
168	$2p_{3/2}6h_{9/2}$	6^{e}	1.97630E+03	---	---	---	---	---	---
169	$2p_{3/2}6g_{9/2}$	6°	1.97631E+03	---	---	---	---	---	---
170	$2p_{3/2}6g_{9/2}$	3°	1.97631E+03	---	---	---	---	---	---
171	$2p_{3/2}6h_{11/2}$	7^{e}	1.97633E+03	---	---	---	---	---	---
172	$2p_{3/2}6h_{9/2}$	4^{e}	1.97635E+03	---	---	---	---	---	---
173	$2p_{3/2}6h_{9/2}$	5^{e}	1.97637E+03	---	---	---	---	---	---
174	$2p_{3/2}6g_{7/2}$	3°	1.97638E+03	---	---	---	---	---	---
175	$2p_{3/2}6h_{11/2}$	5^{e}	1.97638E+03	---	---	---	---	---	---
176	$2p_{3/2}6g_{7/2}$	4°	1.97639E+03	---	---	---	---	---	---
177	$2p_{3/2}6g_{7/2}$	6^{e}	1.97641E+03	---	---	---	---	---	---
178	$2p_{3/2}6g_{9/2}$	4°	1.97642E+03	---	---	---	---	---	---
179	$2p_{3/2}6g_{9/2}$	5°	1.97644E+03	---	---	---	---	---	---
180	$2p_{1/2}6s_{1/2}$	0°	1.99290E+03	---	---	---	---	---	---
181	$2p_{1/2}6s_{1/2}$	1°	1.99302E+03	---	---	---	---	---	---

182	$2p_{1/2}6p_{1/2}$	1^e	1.99809E+03	---	---	---	---	---	---
183	$2p_{1/2}6p_{3/2}$	1^e	1.99900E+03	---	---	---	---	---	---
184	$2p_{1/2}6p_{3/2}$	2^e	1.99910E+03	---	---	---	---	---	---
185	$2p_{1/2}6p_{1/2}$	0^e	1.99929E+03	---	---	---	---	---	---
186	$2p_{1/2}6d_{3/2}$	2°	2.00505E+03	---	---	---	---	---	---
187	$2p_{1/2}6d_{5/2}$	2°	2.00526E+03	---	---	---	---	---	---
188	$2p_{1/2}6d_{5/2}$	3°	2.00538E+03	---	---	---	---	---	---
189	$2p_{1/2}6d_{3/2}$	1°	2.00585E+03	---	2.00879E+03	---	---	---	---
190	$2p_{1/2}6f_{5/2}$	3^e	2.00844E+03	---	---	---	---	---	---
191	$2p_{1/2}6f_{7/2}$	3^e	2.00854E+03	---	---	---	---	---	---
192	$2p_{1/2}6f_{7/2}$	4^e	2.00854E+03	---	---	---	---	---	---
193	$2p_{1/2}6f_{5/2}$	2^e	2.00855E+03	---	---	---	---	---	---
194	$2p_{1/2}6g_{7/2}$	4°	2.00891E+03	---	---	---	---	---	---
195	$2p_{1/2}6h_{9/2}$	4^e	2.00892E+03	---	---	---	---	---	---
196	$2p_{1/2}6s_{g7/2}$	3°	2.00893E+03	---	---	---	---	---	---
197	$2p_{1/2}6h_{9/2}$	5^e	2.00894E+03	---	---	---	---	---	---
198	$2p_{1/2}6h_{11/2}$	5^e	2.00896E+03	---	---	---	---	---	---
199	$2p_{1/2}6g_{9/2}$	5°	2.00896E+03	---	---	---	---	---	---
200	$2p_{1/2}6h_{11/2}$	6^e	2.00897E+03	---	---	---	---	---	---
201	$2p_{1/2}6g_{9/2}$	4°	2.00897E+03	---	---	---	---	---	---
202	$2s_{1/2}5s_{1/2}$	1^e	2.05136E+03	---	---	---	---	---	---
203	$2s_{1/2}5s_{1/2}$	0^e	2.05274E+03	---	---	---	---	---	---
204	$2s_{1/2}5p_{1/2}$	0°	2.06056E+03	---	---	---	---	---	---

205	$2s_{1/2}5p_{1/2}$	1°	2.06068E+03	---	---	---	---	---	---
206	$2s_{1/2}5p_{3/2}$	2°	2.06209E+03	---	---	---	---	---	---
207	$2s_{1/2}5p_{3/2}$	1°	2.06247E+03	---	---	---	---	---	---
208	$2s_{1/2}5d_{3/2}$	1^{e}	2.07254E+03	---	---	---	---	---	---
209	$2s_{1/2}5d_{3/2}$	2^{e}	2.07264E+03	---	---	---	---	---	---
210	$2s_{1/2}5d_{5/2}$	3^{e}	2.07286E+03	---	---	---	---	---	---
211	$2s_{1/2}5d_{5/2}$	2^{e}	2.07389E+03	---	---	---	---	---	---
212	$2s_{1/2}5f_{5/2}$	2°	2.07841E+03	---	---	---	---	---	---
213	$2s_{1/2}5f_{5/2}$	3°	2.07844E+03	---	---	---	---	---	---
214	$2s_{1/2}5f_{7/2}$	4°	2.07853E+03	---	---	---	---	---	---
215	$2s_{1/2}5f_{7/2}$	3°	2.07871E+03	---	---	---	---	---	---
216	$2s_{1/2}5g_{7/2}$	4^{e}	2.07901E+03	---	---	---	---	---	---
217	$2s_{1/2}5g_{7/2}$	3^{e}	2.07901E+03	---	---	---	---	---	---
218	$2s_{1/2}5g_{9/2}$	5^{e}	2.07909E+03	---	---	---	---	---	---
219	$2s_{1/2}5g_{9/2}$	4^{e}	2.07909E+03	---	---	---	---	---	---
220	$2s_{1/2}6s_{1/2}$	1^{e}	2.15120E+03	---	---	---	---	---	---
221	$2s_{1/2}6s_{1/2}$	0^{e}	2.15196E+03	---	---	---	---	---	---

222	$2s_{1/2}6p_{1/2}$	0°	2.15644E+03	---	---	---	---	---	---
223	$2s_{1/2}6p_{1/2}$	1°	2.15652E+03	---	---	---	---	---	---
224	$2s_{1/2}6p_{3/2}$	2°	2.15735E+03	---	---	---	---	---	---
225	$2s_{1/2}6p_{3/2}$	1°	2.15755E+03	---	---	---	---	---	---
226	$2s_{1/2}6d_{3/2}$	1^e	2.16329E+03	---	---	---	---	---	---
227	$2s_{1/2}6d_{3/2}$	2^e	2.16336E+03	---	---	---	---	---	---
228	$2s_{1/2}6d_{5/2}$	3^e	2.16349E+03	---	---	---	---	---	---
229	$2s_{1/2}6d_{5/2}$	2^e	2.16403E+03	---	---	---	---	---	---
230	$2s_{1/2}6f_{5/2}$	2°	2.16658E+03	---	---	---	---	---	---
231	$2s_{1/2}6f_{5/2}$	3°	2.16660E+03	---	---	---	---	---	---
232	$2s_{1/2}6f_{7/2}$	4°	2.16665E+03	---	---	---	---	---	---
233	$2s_{1/2}6f_{7/2}$	3°	2.16676E+03	---	---	---	---	---	---
234	$2s_{1/2}6g_{7/2}$	3^e	2.16703E+03	---	---	---	---	---	---
235	$2s_{1/2}6g_{7/2}$	4^e	2.16703E+03	---	---	---	---	---	---
236	$2s_{1/2}6h_{9/2}$	5°	2.16705E+03	---	---	---	---	---	---
237	$2s_{1/2}6h_{9/2}$	4°	2.16705E+03	---	---	---	---	---	---
238	$2s_{1/2}6g_{9/2}$	5^e	2.16708E+03	---	---	---	---	---	---
239	$2s_{1/2}6h_{11/2}$	6°	2.16708E+03	---	---	---	---	---	---
240	$2s_{1/2}6h_{11/2}$	5°	2.16708E+03	---	---	---	---	---	---
241	$2s_{1/2}6g_{9/2}$	4^e	2.16708E+03	---	---	---	---	---	---

Index: Level index. **State Configuration**: The configuration to which the level belongs (jj-coupling). $\mathbf{J^{\pi}}$: J is the Total angular momentum, π is the Parity. **FAC**: Energies from the FAC code (our calculation). **COWAN**: Energies from the COWAN code (also calculated). **NIST**: National Institute of Standards and Technology, http://www.nist.gov [24]. **Exp.**: Energies calculated experimentally [23]. **CIV3**: Energies from the CIV3 code [22]. **MCDF**: Energies from the MCDF code [23]. **RCI**: Energies from the RCI code [23].

In this paper, we calculate energy levels for 241 fine-structure states using a fully relativistic approach based on Dirac equation (see Table 1). Weighted Oscillator strengths, spontaneous radiative decay rates are calculated in the single multipole approximation (see Table 2), and Collision strengths by electron impact using the factorization-interpolation method are calculated in the distorted wave approximation. Effective collision strengths are calculated by interpolating the data from the collision strengths and integrating over Max wellian distribution at different temperatures. Rate coefficients are calculated from effective collision strengths using a formula that will be described later in this paper. Then, we predict the reduced population and gain coefficient for Ge XXIII by a steady state equation in the collisional radiative model after achieving a population inversion between the allowed transition states.

Index: Level index. State Configuration: The configuration to which the level belongs (jj-coupling). Jπ: J is the Total angular momentum, π is the Parity. FAC: Energies from the FAC code (our calculation). COWAN: Energies from the COWAN code (also calculated). NIST: National Institute of Standards and Technology, http://www.nist.gov [24] . Exp.: Energies calculated experimentally [23] . CIV3: Energies from the CIV3 code [22] . MCDF: Energies from the MCDF code [23] . RCI: Energies from the RCI code [23] .

i and j: The lower (i) and upper (j) levels of a transition as defined in Table 4. gf (nm) (FAC): Weighted oscillator strengths from the FAC code (our calculation). gf (nm) (COWAN): Weighted oscillator strengths from the COWAN code (also calculation). gf (nm) (CIV3): Weighted oscillator strengths from the CIV3 code [22] . gf (nm) (RCI): Weighted oscillator strengths from the RCI code [23] . Aji (s−1) (FAC): Transition probabilities (in nm) from the FAC code (our calculation). Aji (s−1) (COWAN): Transition probabilities (in nm) from the COWAN code (also calculation). Aji (s−1) (CIV3): Transition probabilities (in nm) from the CIV3 code [22] .

Table 2. Comparison between some weighted oscillator strengths, and transition probabilities for E1 transitions of Ge XXIII.

j	i	gf (FAC)	gf (COWAN)	gf (CIV3)	gf (RCI)	A_{ji} (s^{-1}) (FAC)	A_{ji} (s^{-1}) (COWAN)	A_{ji} (s^{-1}) (CIV3)
3	1	1.29E-01	1.0139E-01	1.280E-01	----	2.86E+12	2.2503E+12	2.84E+12
5	1	8.88E-02	7.0307E-02	8.480E-02	----	2.07E+12	1.6400E+12	1.1868E+12
6	2	2.26E-01	2.5061E-01	2.257E-01	0.2685	5.96E+09	6.5800E+09	5.9267E+09
7	2	2.13E-01	2.1330E-01	2.113E-01	0.2246	3.96E+09	3.9660E+09	3.9080E+09
7	3	2.15E-01	2.2961E-01	2.185E-01	0.2204	3.51E+09	3.7720E+09	3.5940E+09
8	2	7.00E-01	7.2277E-01	7.002E-01	0.7275	1.18E+10	1.2177E+10	1.1760E+10
9	2	4.53E-04	1.1885E-04	4.000E-04	0.0001	1.85E+07	4.7833E+06	1.5900E+07
9	3	2.84E-01	2.9512E-01	2.827E-01	0.2914	1.03E+10	1.0657E+10	1.0260E+10
10	2	2.91E-01	2.9107E-01	2.923E-01	0.3092	8.02E+09	7.9700E+09	7.9360E+09
10	3	2.52E-01	2.4604E-01	2.480E-01	0.2431	6.27E+09	6.0900E+09	6.1160E+09
10	5	6.21E-04	8.1658E-04	5.000E-04	----	2.70E+06	3.4520E+06	2.3540E+06
11	3	1.08E-01	1.0495E-01	1.102E-01	0.0930	1.98E+10	1.9050E+10	2.0180E+10
11	5	1.26E-02	1.1588E-02	1.130E-02	----	6.31E+08	5.5900E+08	5.6060E+08
12	2	6.08E-04	4.2267E-04	7.000E-04	0.0000	5.31E+07	3.6867E+07	6.0800E+07
12	3	5.51E-04	2.8249E-04	7.000E-04	0.0076	4.46E+07	2.2927E+07	5.2800E+07

12	4	8.97E−02	9.1622E−02	8.920E−02	0.1111	2.65E+09	2.7023E+09	2.6460E+09
12	5	1.59E−01	1.6943E−01	1.601E−01	0.1542	4.37E+09	4.6533E+09	4.4367E+09
13	2	1.24E−02	6.9183E−03	1.300E−02	----	1.31E+09	7.3167E+08	1.3680E+09
13	3	2.86E−04	5.2481E−05	4.000E−04	----	2.83E+07	5.1933E+06	3.5600E+07
13	4	1.96E−01	2.0417E−01	1.952E−01	0.2017	8.02E+09	8.3033E+09	7.9133E+09
13	5	9.70E−02	1.0139E−01	9.450E−02	0.1224	3.74E+09	3.8800E+09	3.6167E+09
14	2	1.78E−03	1.2823E−03	1.800E−03	----	1.16E+08	8.3220E+07	1.1690E+08
14	3	1.08E−03	1.2853E−03	1.000E−03	----	6.57E+07	7.8180E+07	6.2340E+07
14	5	5.08E−01	5.1642E−01	5.074E−01	0.5590	1.22E+10	1.2278E+10	1.2140E+10
15	3	5.19E−02	3.7154E−02	4.470E−02	----	2.14E+10	1.4450E+10	1.7370E+10
15	5	1.15E−01	1.0447E−01	1.125E−01	0.1002	2.19E+10	1.8200E+10	1.9640E+10
16	6	1.02E−01	1.1015E−01	9.950E−02	0.1190	1.80E+10	1.9500E+10	1.7480E+10
16	13	8.34E−04	7.3961E−04	8.000E−04	----	1.57E+07	1.4110E+07	1.5230E+07
17	1	9.08E−03	1.7539E−02	1.070E−02	----	2.37E+11	4.5833E+11	2.7840E+11
17	6	2.41E−01	2.5061E−01	2.368E−01	0.1374	1.50E+10	1.5670E+10	1.4670E+10
17	7	2.75E−02	3.1477E−02	2.580E−02	0.0416	1.53E+09	1.7543E+09	1.4320E+09
17	9	1.03E−03	2.2961E−03	1.400E−03	0.0578	4.56E+07	1.0280E+08	6.0967E+07
17	10	3.59E−02	3.9902E−02	3.570E−02	0.0095	1.40E+09	1.5723E+09	1.4040E+09
17	11	3.27E−03	5.1168E−03	3.300E−03	0.0551	7.69E+07	1.2400E+08	7.9167E+07
18	8	9.17E−01	2.3227E−02	1.880E−02	----	1.54E+10	7.1220E+08	3.1800E+08
19	6	1.78E−01	----	1.692E-01	0.1613	7.31E+09	----	3.8478E+09

19	7	7.27E−02	----	----	0.0483	2.69E+09	----	----
19	8	2.05E−02	9.4842E−01	9.122E−01	0.0245	6.25E+08	1.6189E+10	2.7820E+10
19	9	5.28E−02	----	----	0.0966	1.57E+09	----	----
19	10	1.94E−01	----	----	0.2101	5.12E+09	----	----
19	12	2.59E−04	----	----	----	2.58E+06	----	----
19	14	5.61E−04	----	----	----	3.02E+06	----	----
20	7	6.62E−01	6.8707E−01	6.600E−01	0.6821	1.76E+10	1.8343E+10	1.7643E+10
20	8	1.08E−01	8.6896E−02	1.058E−01	0.1138	2.37E+09	1.9229E+09	2.3443E+09
20	10	9.78E−03	2.7102E−02	1.010E−02	0.0087	1.86E+08	5.2129E+08	1.9514E+08
21	6	2.98E−02	5.4325E−02	3.670E−02	0.0597	1.34E+09	2.4340E+09	1.6360E+09
21	7	1.75E−01	2.0370E−01	----	0.2015	7.08E+09	8.2260E+09	7.3560E+09
21	8	6.42E−03	9.7051E−03	----	0.0118	2.16E+08	3.2740E+08	2.4860E+08
21	9	3.47E−01	3.0761E−01	----	0.3133	1.14E+10	1.0172E+10	1.0940E+10
21	10	6.08E−03	3.9355E−05	----	0.0005	1.78E+08	1.1614E+06	9.4940E+07
22	7	3.17E−03	2.3659E−04	----	0.1574	9.73E+07	7.2943E+06	8.9543E+07
22	8	1.51E−01	1.7219E−01	----	0.0045	3.87E+09	4.4671E+09	3.8886E+09
22	10	6.04E−01	5.9429E−01	----	0.6387	1.36E+10	1.3539E+10	1.3590E+10
23	1	1.21E+00	1.4060E+00	----	----	3.23E+13	3.7600E+13	3.2120E+13
23	9	1.59E−01	1.5560E−01	----	0.0691	1.19E+10	1.1437E+10	1.1690E+10

24	10	3.70E−04	7.4473E−04	----	----	2.46E+07	4.9180E+07	3.8180E+07
24	12	4.79E−01	4.7643E−01	----	0.0191	1.83E+10	1.7818E+10	1.7880E+10
24	13	1.13E−02	1.6943E−05	----	----	3.34E+08	4.9140E+05	2.1240E+08
24	14	6.99E−02	9.1201E−02	----	0.0608	1.99E+09	2.5640E+09	2.0880E+09
25	6	9.36E−04	6.1235E−04	----	----	8.59E+07	5.5860E+07	6.6320E+07
25	7	2.24E−03	3.6141E−03	----	----	1.91E+08	3.0660E+08	2.0400E+08
25	8	3.72E−03	1.3428E−03	----	----	2.81E+08	1.0084E+08	3.4300E+08
25	9	1.27E−04	4.6345E−06	----	----	9.45E+06	3.4380E+05	1.2870E+07
25	10	8.28E−03	6.5615E−03	----	----	5.71E+08	4.5180E+08	6.4040E+08
25	12	2.48E−04	9.8628E−03	----	0.0028	9.91E+06	3.8980E+08	1.7430E+06
25	13	4.55E−01	4.7863E−01	----	0.4467	1.41E+10	1.4806E+10	1.4020E+10
25	14	5.15E−02	3.3651E−02	----	0.0653	1.55E+09	1.0066E+09	1.3530E+09
26	7	2.06E−04	4.4157E−05	----	----	1.29E+07	2.7300E+06	1.5157E+07
26	8	2.27E−03	2.3174E−03	----	----	1.26E+08	1.2674E+08	1.3200E+08
26	10	1.87E−04	7.6384E−04	----	----	9.45E+06	3.8286E+07	1.0320E+07
26	14	7.32E−01	7.4131E−01	----	0.7396	1.63E+10	1.6329E+10	1.6200E+10
27	1	2.14E+00	1.9861E+00	----	----	5.97E+13	5.5400E+13	5.9600E+13
27	12	1.11E−01	1.1041E−01	----	0.0948	9.62E+09	9.0600E+09	9.1033E+09
27	15	1.44E−01	1.5524E−01	----	0.1571	6.28E+09	6.9133E+09	6.8567E−09
28	2	3.12E−01	----	----	----	1.70E+11	----	9.3233E+10

28	3	7.53E−02	----	----	----	3.97E+10	----	2.1890E+10
28	4	5.48E−02	----	----	----	2.07E+10	----	1.0920E+10
28	5	9.63E−02	----	----	----	3.56E+10	----	1.8670E+10
29	3	7.57E−02	----	----	----	1.32E+11	----	6.7650E+10
29	5	3.91E−02	----	----	----	4.86E+10	----	2.1740E+10
29	23	2.05E−03	----	----	----	5.43E+08	----	1.3400E+08
29	27	3.36E−03	----	----	0.0003	3.38E+08	----	2.6180E+07
30	6	7.81E−02	----	----	----	1.30E+11	----	6.2310E+10
30	9	3.01E−02	----	----	----	4.51E+10	----	2.1320E+10
30	12	5.20E−02	----	----	----	5.86E+10	----	2.4570E+10
30	13	1.28E−03	----	----	----	1.31E+09	----	9.2910E+08
30	28	9.37E−02	----	----	0.0866	8.21E+09	----	4.9130E+09
31	1	6.68E−02	----	----	----	2.10E+12	----	1.6030E+12
31	6	4.63E−02	----	----	----	2.60E+10	----	1.2140E+10
31	7	2.89E−01	----	2.054E−01	----	1.56E+11	----	7.3700E+10
31	9	7.46E−04	----	1.200E−03	----	3.75E+08	----	3.9933E+08
31	10	4.70E−03	----	4.100E−03	----	2.28E+09	----	1.2790E+09
31	11	1.73E−02	----	1.140E−02	----	7.37E+09	----	3.0550E+09
31	12	9.38E−02	----	6.860E−02	----	3.56E+10	----	1.5760E+10
31	13	4.43E−03	----	3.700E−03	----	1.52E+09	----	7.5467E+08
31	14	2.32E−03	----	1.200E−03	----	7.86E+08	----	2.2760E+08

31	29	3.84E−02	----	3.270E−02	----	7.34E+08	----	4.5700E+08
32	6	5.93E−03	----	2.600E−03	----	2.14E+09	----	6.4740E+08
32	7	1.59E−03	----	9.000E−04	----	5.54E+08	----	2.1140E+08
32	8	4.35E−01	----	3.034E−01	----	1.42E+11	----	6.6100E+10
32	9	2.05E−02	----	1.370E−02	----	6.65E+09	----	2.9480E+09
32	10	1.28E−01	----	9.580E−02	----	4.02E+10	----	1.9850E+10
32	12	9.76E−04	----	8.000E−04	----	2.41E+08	----	1.1380E+08
32	13	8.13E−02	----	5.510E−02	----	1.82E+10	----	7.4940E+09
32	14	1.20E−01	----	8.710E−02	----	2.65E+10	----	1.1660E+10
32	28	5.61E−01	----	4.884E−01	----	1.35E+10	----	8.8360E+09
33	1	2.94E−01	----	2.689E−01	----	9.37E+12	----	8.1633E+12
33	6	6.25E−03	----	4.300E−03	----	3.85E+09	----	1.7920E+09
33	7	6.11E−03	----	4.400E−03	----	3.63E+09	----	1.7370E+09
33	9	1.18E−01	----	8.710E−02	----	6.55E+10	----	3.2250E+10
33	10	1.36E−01	----	9.750E−02	----	7.27E+10	----	3.4667E+10
33	11	3.92E−02	----	3.350E−02	----	1.86E+10	----	1.0200E+10
33	12	8.19E−03	----	7.400E−03	----	3.48E+09	----	1.9470E+09
33	13	3.96E−02	----	3.140E−02	----	1.53E+10	----	7.3833E+09
33	14	1.19E−01	----	8.090E−02	----	4.54E+10	----	1.8700E+10
33	15	4.86E−02	----	3.990E−02	----	1.55E+10	----	7.6933E+09
33	28	6.38E−02	----	5.630E−02	0.0697	2.80E+09	----	1.8650E+09
33	29	2.21E−01	----	1.892E−01	0.2430	6.69E+09	----	4.4000E+09
34	16	2.63E−02	----	1.660E−02	----	1.46E+10	----	5.9533E+09
34	17	6.04E−02	----	3.970E−02	----	3.30E+10	----	1.3880E+10

34	19	3.26E−02	----	----	----	1.72E+10	----	----
34	21	1.13E−01	----	8.290E−02	----	5.80E+10	----	2.6690E+10
34	23	4.12E−02	----	2.970E−02	----	1.87E+10	----	8.2633E+09
34	24	1.09E−01	----	7.540E−02	----	3.97E+10	----	1.5690E+10
34	25	8.82E−04	----	2.000E−04	----	3.15E+08	----	5.5600E+07
34	27	1.63E−02	----	1.250E−02	----	5.13E+09	----	2.1360E+09
34	30	1.98E−01	----	1.875E−01	0.2028	1.16E+10	----	1.0040E+10
34	31	1.19E−01	----	1.107E−01	0.1190	6.80E+09	----	5.7400E+09
34	32	8.58E−03	----	8.000E−03	----	3.91E+08	----	3.2560E+08
34	33	2.28E−02	----	2.230E−02	0.0207	9.49E+08	----	8.3367E+08
35	17	2.53E−02	----	1.570E−02	----	8.32E+09	----	3.3160E+09
35	19	9.06E−02	----	----	----	2.88E+10	----	----
35	20	3.05E−01	----	2.106E−01	----	9.68E+10	----	4.2360E+10
35	21	5.55E−03	----	----	----	1.71E+09	----	9.6780E+08
35	22	3.89E−02	----	----	----	1.17E+10	----	5.5680E+09
35	23	3.11E−02	----	----	----	8.53E+09	----	3.4440E+09
35	24	7.59E−02	----	----	----	1.67E+10	----	6.0480E+09
35	25	3.18E−02	----	----	----	6.85E+09	----	3.3620E+09
35	26	4.46E−02	----	----	----	9.47E+09	----	3.8180E+09

35	31	3.98E−01	----	----	0.4009	1.39E+10	----	1.1860E+10
35	32	1.30E−01	----	----	----	3.60E+09	----	3.0320E+09
35	33	4.04E−02	----	----	0.0403	1.02E+09	----	9.4340E+08
36	18	5.29E−01	----	----	----	1.22E+11	----	2.4014E+09
36	19	2.52E−02	----	----	----	5.79E+09	----	5.4543E+10
36	20	5.20E−04	----	----	----	1.19E+08	----	7.0343E+07
36	21	1.27E−02	----	----	----	2.82E+09	----	1.0570E+09
36	22	1.65E−01	----	----	----	3.59E+10	----	1.5700E+10
36	24	1.53E−03	----	----	----	2.44E+08	----	7.1843E+07
36	25	8.08E−02	----	----	----	1.26E+10	----	3.9943E+09
36	26	1.33E−01	----	----	----	2.04E+10	----	7.9014E+09
36	32	7.51E−01	----	----	----	1.54E+10	----	1.3070E+10
37	3	6.24E−04	----	----	----	5.14E+08	----	2.8880E+08
37	5	5.56E−04	----	----	----	3.70E+08	----	2.3160E+08
37	17	1.26E−03	----	----	----	4.51E+08	----	2.2120E+08
37	19	2.48E−02	----	----	----	8.57E+09	----	----
37	20	1.04E−01	----	----	----	3.58E+10	----	1.7050E+10
37	21	1.20E−01	----	----	----	4.03E+10	----	1.7640E+10
37	22	1.48E−01	----	----	----	4.87E+10	----	2.1580E+10
37	23	3.73E−02	----	----	----	1.12E+10	----	4.3680E+09
37	24	3.98E−02	----	----	----	9.64E+09	----	4.2460E+09
37	25	4.45E−02	----	----	----	1.06E+10	----	3.8760E+09
37	26	1.42E−01	----	----	----	3.34E+10	----	1.3290E+10
37	27	3.24E−02	----	----	----	6.84E+09	----	1.6950E+09

37	31	8.53E−02	----	----	0.0855	3.78E+09	----	3.5220E+09
37	32	1.24E−03	----	----	0.0019	4.49E+07	----	2.9640E+07
37	33	5.14E−01	----	----	0.5069	1.72E+10	----	1.4760E+10
38	6	1.56E−01	1.6368E−01	----	----	2.09E+11	2.2080E+11	----
38	7	1.23E−01	1.3274E−01	----	----	1.62E+11	1.7580E+11	----
38	8	3.78E−01	3.8548E−01	----	----	4.83E+11	4.9520E+11	----
38	10	1.55E−01	1.4859E−01	----	----	1.94E+11	1.8694E+11	----
38	12	1.32E−04	1.7701E−04	----	----	1.47E+08	1.9826E+08	----
38	13	3.36E−03	3.0130E−03	----	----	3.57E+09	3.2240E+09	----
38	14	9.20E−04	5.7810E−04	----	----	9.74E+08	6.1580E+08	----

i and j: The lower (i) and upper (j) levels of a transition as defined in Table 4. **gf (nm) (FAC)**: Weighted oscillator strengths from the FAC code (our calculation). **gf (nm) (COWAN)**: Weighted oscillator strengths from the COWAN code (also calculation). **gf (nm) (CIV3)**: Weighted oscillator strengths from the CIV3 code [22]. **gf (nm) (RCI)**: Weighted oscillator strengths from the RCI code [23]. **A_{ji} (s^{-1}) (FAC)**: Transition probabilities (in nm) from the FAC code (our calculation). **A_{ji} (s^{-1}) (COWAN)**: Transition probabilities (in nm) from the COWAN code (also calculation). **A_{ji} (s^{-1}) (CIV3)**: Transition probabilities (in nm) from the CIV3 code [22].

2. GAIN COEFFICIENT COMPUTATIONS

The possibility of laser emission from plasma of Ge XXIII ion via electron collisional pumping, in the XUV and soft X-ray spectral regions is investigated at different plasma temperatures and plasma electron densities. The reduced population densities are calculated by solving the coupled rate equations [13] [14] .

$$N_j\left[\sum_{i<j}A_{ji}+N_e\left(\sum_{i<j}C_{ji}^d+\sum_{i>j}C_{ji}^e\right)\right]=N_e\left(\sum_{i<j}N_iC_{ij}^e+\sum_{i>j}N_iC_{ij}^d\right)+\sum_{i>j}N_iA_{ij} \tag{1}$$

where N_j and N_i is the fractional population of level j and i respectively, N_e is the electron density, A_{ji} is the Einstein coefficient for spontaneous radiative decay from j to i; and C_{ij}^e and C_{ji}^d represent the rate coefficient for collisional excitation and de-excitation respectively. The actual population density N_J of the jth level can be calculated from the equation of identity [15] [16] .

$$C_{ji}^d=C_{ij}^e\left[\frac{g_i}{g_j}\right]\exp\left[\frac{\Delta E_{ji}}{kT_e}\right] \tag{2}$$

where g_i and g_j are the statistical weights of the lower and upper levels, respectively.

The electron impact excitation rates usually are expressed via the effective collision strengths γ_{ji} as

$$C_{ji}^d=\frac{8.6287\times10^{-6}}{g_jT_e^{1/2}}\gamma_{ij} \tag{3}$$

where the values of γ_{ji} and A_{ji} are obtained by [17] .

The actual population density N_J of the jth level is obtained from the following identity [17] ,

$$N_J = N_j \times N_i \tag{4}$$

where N_I is the quantity of ions which reached to the ionization stage I [17] ,

$$N_I = f_I \, N_e / Z_{avg} \tag{5}$$

where N_e is the electron density, Z_{avg} is the average degree of ionization and f_I is the fractional abundance of the ionization states which can be calculated from the relation [17] . Since the populations calculated from Equation (1) are normalized such that,

$$\sum_{J=1}^{241} \frac{N_J}{N_I} = 1 \tag{6}$$

where 241 is the number of all the levels of the ion under consideration, the quantity actually obtained from Equation (1) is the fractional population N_J/N_I. After the calculation of levels population, the quantities N_j/g_j and N_i/g_i can be calculated. Application of electron collisional pumping, the collision in the laser ion plasma will transfer the pumped quanta to other levels, and will result in population inversions between the upper and lower levels. Once a population inversion has been ensured a positive gain through $F > 0$ [18] is obtained.

$$F = \frac{g_j}{N_j}\left[\frac{N_j}{g_j} - \frac{N_i}{g_i}\right] \tag{7}$$

where N_j/g_j and N_i/g_i are the reduced populations of the upper level and lower level respectively. Equation (7) has been used to calculate the gain coefficient (α) for Doppler broadening of the various transitions in the Ge XXIII ion.

$$\alpha_{ji} = \frac{\lambda_{ij}^3}{8\pi}\left[\frac{M}{2\pi KT_i}\right]^{1/2} A_{ji} N_j F \tag{8}$$

where M is the ion mass λ_{ij} is the transition wavelength in cm (see Table 3), T_i is the ion temperature in K and j, i represents the upper and lower transition levels respectively.

As seen from Equation (8), the gain coefficient is expressed in terms of the upper state density (N_j). This quantity N_j depends on how the upper state is populated, as well as on the density of the initial source state. The source state is often the ground state for the particular ion.

3. RESULTS AND DISCUSSIONS

3.1. Energy Levels

The energy level values obtained using the fully relativistic flexible atomic code (FAC) for the 1s2 2s2 2p5nl, 1s2 i and j: The lower (i) and upper (j) levels of a transition as defined in Table 4. λ (nm) (FAC): Transition wavelengths (in nm) from the FAC code (our calculation). λ (nm)(COWAN): Transition wavelengths (in nm) from the COWAN code (also calculation). λ (nm) (CIV3): Transition wavelengths (in nm) from the CIV3 code [22] . λ (nm) (MCDF): Transition wavelengths (in nm) from the MCDF code [23] . λ (nm) (RCI): Transition wavelengths (in nm) from the RCI code [23] . λ (nm) (Exp.): Transition wavelengths (in nm) calculated experimentally [23] .

2s1 2p6nl (n = 3, 4, 5, 6; l = s, p, d, f, g and h) configurations in Ne-like Ge ions are presented in Table 1. The main components of the computed Eigen vectors are also given in these tables in the jj-coupling schemes. For Ge XXIII, the agreement between FAC, MCDF, and other theoretical and experimental energies [19] - [23] with the values available at the National Institute of Standards and Technology (NIST) [24] and is within values less than 0.5% for a majority of levels.

Table 3. Comparison between some wavelengths (λ) for E1 transitions of Ge XXIII (in nm).

j	i	λ (nm) (FAC)	λ (nm) (COWAN)	λ (nm) (CIV3)	λ (nm) (MCDF)	λ (nm) (RCI)	λ (nm) (Exp.)
3	1	1.00291911	1.0009	1.001	----	----	----
5	1	0.978298138	0.9763	0.976	----	----	----
6	2	29.06054601	29.0804	29.1	----	----	----
7	2	26.82759761	26.7880	26.85	----	----	----
7	3	28.60854399	28.5017	28.48	28.644	28.432	28.628
8	2	23.76385089	23.7695	23.82	23.760	24.192	23.767
9	2	23.38618832	23.5053	23.51	23.395	24.151	23.385
9	3	24.72809857	24.8145	24.75	24.727	25.422	24.735
10	2	22.00458606	22.0738	22.16	----	----	----
10	3	23.18861213	23.2245	23.26	23.229	24.033	23.224
10	5	55.46243419	56.1774	56.06	----	----	----
11	3	19.06599826	19.1780	19.09	----	----	----
11	5	36.55639248	37.1945	36.72	----	----	----
12	2	15.97797365	15.9610	15.98	----	----	----
12	3	16.59318561	16.5540	16.54	----	----	----
12	4	27.43670009	27.4431	27.39	----	----	----
12	5	28.43226568	28.4487	28.32	28.475	28.318	28.509
13	2	14.48681287	14.4972	14.54	----	----	----
13	3	14.99074178	14.9848	15	----	----	----
13	4	23.31563846	23.3836	23.42	23.319	23.390	23.314
13	5	24.03069931	24.1098	24.1	24.044	23.900	24.049
14	2	14.32320141	14.3395	14.36	----	----	----
14	3	14.81561859	14.8164	14.81	----	----	----

14	5	23.58382453	23.6768	23.61	23.627	23.407	23.625
15	3	12.74125872	13.0918	13.1	----	----	----
15	5	18.72982602	19.5593	19.55	19.544	19.399	19.619
16	6	19.4465226	19.4149	19.49	19.399	19.701	19.460
16	13	59.50360748	59.1173	59.16	----	----	----
17	1	0.924189369	0.9223	0.922	----	----	----
17	6	18.92164195	18.8596	18.94	18.878	19.045	18.940
17	7	20.00584033	19.9678	20.03	----	----	----
17	9	22.47182417	22.2880	22.41	----	----	----
17	10	23.91464589	23.7483	23.78	----	----	----
17	11	30.77812808	30.2817	30.63	----	----	----
18	8	20.99113251	20.8451	20.93	----	----	----
19	6	18.02121903	----	18.05	----	----	----
19	7	19.00201092	----	----	18.895	18.877	18.961
19	8	20.91160561	20.8281	20.91	----	----	----
19	9	21.21305773	----	----	----	----	----
19	10	22.49415888	----	----	----	----	----
19	12	36.61005733	----	----	----	----	----
19	14	49.79022451	----	----	----	----	----

20	7	18.94688793	18.9013	18.88	18.814	18.874	18.880
20	8	20.84486629	20.7616	20.74	----	----	----
20	10	22.4169586	22.2548	22.18	----	----	----
21	6	17.27924434	17.2601	17.3	----	----	----
21	7	18.17892	18.1837	18.2	----	----	----
21	8	19.9190906	19.8990	19.92	19.814	19.669	19.873
21	9	20.1924195	20.0880	20.14	----	----	----
21	10	21.3498462	21.2667	21.25	----	----	----
22	7	17.6454186	17.5769	17.67	----	----	----
22	8	19.2803566	19.1747	19.29	19.192	18.936	19.259
22	10	20.617748	20.4414	20.53	20.475	19.815	20.539
23	1.	0.91269295	0.9117	0.911	----	----	----
23	9	17.2029364	17.3848	17.21	----	----	----
23	11	21.6825457	21.8925	21.67	----	----	----
24	6	12.2585873	12.3094	12.28	----	----	----
24	7	12.704653	12.7720	12.73	----	----	----
24	10	14.1760934	14.2199	14.14	----	----	----
24	12	18.7265184	18.8773	18.79	----	----	----
24	13	21.295593	21.4373	21.26	----	----	----
24	14	21.6592895	21.7917	21.65	----	----	----
25	6	12.0678472	12.0887	12.1	----	----	----
25	7	12.4998937	12.5346	12.54	----	----	----
25	8	13.2987553	13.3265	13.33	----	----	----
25	9	13.4200377	13.4110	13.43	----	----	----

25	10	13.921634	13.9263	13.91	----	----	----
25	12	18.285025	18.3634	18.38	----	----	----
25	13	20.7264947	20.7769	20.74	----	----	----
25	14	21.0708515	21.1096	21.11	----	----	----
26	7	12.3577809	12.4198	12.38	----	----	----
26	8	13.138017	13.1968	13.15	----	----	----
26	10	13.7455841	13.7847	13.71	----	----	----
26	14	20.6701616	20.7860	20.65	20.519	20.644	20.583
27	1	0.89303659	0.8927	0.892	----	----	----
27	12	16.0204352	16.4624	16.1	----	----	----
27	15	22.6240255	22.3370	21.61	----	----	----
28	2	6.39977902	----	7.52	----	----	----
28	3	6.49625131	----	7.64	----	----	----
28	4	7.68540454	----	9.36	----	----	----
28	5	7.76153141	----	9.46	----	----	----
29	3	6.19256802	----	7.29	----	----	----
29	5	7.33194183	----	8.93	----	----	----
29	23	15.8945958	----	25.93	----	----	----
29	27	25.774576	----	68.98	----	----	----

30	6	6.32686538	----	7.74	----	----	----
30	9	6.67972533	----	8.27	----	----	----
30	12	7.69935951	----	9.91	----	----	----
30	13	8.10118282	----	10.56	----	----	----
30	28	27.6154168	----	32.75	----	----	----
31	1	0.84183128	----	0.861	----	----	----
31	6	6.30092271	----	7.7	----	----	----
31	7	6.416725578	----	7.87	----	----	----
31	9	6.650814849	----	8.21	----	----	----
31	10	6.77173044	----	8.39	----	----	----
31	11	7.228153142	----	9.11	----	----	----
31	12	7.660979329	----	9.83	----	----	----
31	13	8.058697858	----	10.47	----	----	----
31	14	8.110232992	----	10.57	----	----	----
31	15	8.903756621	----	11.65	----	----	----
31	28	27.1279177	----	31.94	----	----	----
31	29	34.1140454	----	39.92	----	----	----
32	6	6.089149806	----	7.38	----	----	----
32	7	6.197229399	----	7.54	----	----	----
32	8	6.387459838	----	7.82	----	----	----
32	9	6.415305514	----	7.86	----	----	----
32	10	6.527736953	----	8.02	----	----	----
32	12	7.35016756	----	9.33	----	----	----
32	13	7.715504567	----	9.9	----	----	----

32	14	7.762726026	----	9.98	----	----	----
32	28	23.59488012	----	27.15	----	----	----
33	1	0.836515264	----	0.856	----	----	----
33	6	6.014819126	----	7.29	----	----	----
33	7	6.120253309	----	7.44	----	----	----
33	9	6.33285591	----	7.75	----	----	----
33	10	6.442391307	----	7.91	----	----	----
33	11	6.854144356	----	8.54	----	----	----
33	12	7.242135475	----	9.18	----	----	----
33	13	7.596553105	----	9.73	----	----	----
33	14	7.642330131	----	9.81	----	----	----
33	15	8.342972988	----	10.74	----	----	----
33	28	22.51667288	----	25.91	----	----	----
33	29	27.12777534	----	30.91	----	----	----
34	16	6.32759139	----	7.88	----	----	----
34	17	6.385224484	----	7.97	----	----	----
34	19	6.494731238	----	----	----	----	----
34	21	6.596819889	----	8.31	----	----	----
34	23	6.993883236	----	8.94	----	----	----

34	24	7.819488144	----	10.34	----	----	----
34	25	7.89912776	----	10.46	----	----	----
34	27	8.412867128	----	11.38	----	----	----
34	30	19.45339221	----	20.38	----	19.340	19.209
34	31	19.70280794	----	20.71	----	----	----
34	32	22.10702133	----	23.38	----	----	----
34	33	23.14544859	----	24.39	----	----	----
35	17	6.370709831	----	7.95	----	----	----
35	19	6.479715048	----	----	----	----	----
35	20	6.486151251	----	8.144	----	----	----
35	21	6.58132848	----	8.28	----	----	----
35	22	6.65416425	----	8.393	----	----	----
35	23	6.97647331	----	8.9	----	----	----
35	24	7.79773163	----	10.29	----	----	----
35	25	7.87692645	----	10.42	----	----	----
35	26	7.93442318	----	10.53	----	----	----
35	27	8.38768868	----	11.33	----	----	----
35	31	19.5652593	----	20.52	----	19.544	19.414
35	32	21.9340035	----	23.15	----	----	----
35	33	22.9558645	----	24.14	----	----	----
36	18	6.44169557	----	8.07	----	----	----
36	19	6.44922283	----	8.07	----	----	----
36	20	6.45559522	----	8.096	----	----	----
36	21	6.54987129	----	8.23	----	----	----

36	22	6.62200863	----	8.342	----	----	----
36	24	7.75361057	----	10.21	----	----	----
36	25	7.83190723	----	10.34	----	----	----
36	26	7.88874624	----	10.45	----	----	----
36	32	21.5884556	----	22.76	----	----	----
37	3	4.02622122	----	4.6	----	----	----
37	5	4.47873387	----	5.21	----	----	----
37	17	6.11875924	----	7.57	----	----	----
37	19	6.21924501	----	----	----	----	----
37	20	6.22517394	----	7.745	----	----	----
37	21	6.3127944	----	7.87	----	----	----
37	22	6.37977744	----	7.969	----	----	----
37	23	6.6754629	----	8.43	----	----	----
37	24	7.42357701	----	9.66	----	----	----
37	25	7.4953238	----	9.77	----	----	----
37	26	7.54736614	----	9.87	----	----	----
37	27	7.95634814	----	10.57	----	----	----
37	31	17.3688062	----	18.16	----	----	----
37	32	19.2105275	----	20.19	----	----	----

37	33	19.9898721	----	20.94	----	19.959	19.925
38	6	3.15614026	3.1459	----	----	----	----
38	7	3.18493053	3.1753	----	----	----	----
38	8	3.23443596	3.2238	----	----	----	----
38	10	3.27002	3.2577	----	----	----	----
38	12	3.46419347	3.4529	----	----	----	----
38	13	3.5432683	3.5300	----	----	----	----
38	14	3.55319554	3.5395	----	----	----	----

i and j: The lower (i) and upper (j) levels of a transition as defined in Table 4. **λ (nm) (FAC)**: Transition wavelengths (in nm) from the FAC code (our calculation). **λ (nm)(COWAN)**: Transition wavelengths (in nm) from the COWAN code (also calculation). **λ (nm) (CIV3)**: Transition wavelengths (in nm) from the CIV3 code [22]. **λ (nm) (MCDF)**: Transition wavelengths (in nm) from the MCDF code [23]. **λ (nm) (RCI)**: Transition wavelengths (in nm) from the RCI code [23]. **λ (nm) (Exp.)**: Transition wavelengths (in nm) calculated experimentally [23].

3.2. Radiative Decay Rates

The oscillator strengths (f_{ij}) and radiative rates (A_{ij}) (in s^{-1}) for a transition i→ j are related by the following expression

$$f_{ij} = \frac{mc}{8\pi^2 e^2} \lambda_{ij}^2 \frac{\omega_j}{\omega_i} A_{ji} = 1.49\times10^{-16} \lambda_{ij}^2 \frac{\omega_j}{\omega_i} A_{ji}$$

where m and e are the electron mass and charge, respectively, c is the velocity of light, λ_{ij} is the transition wavelength in Å and ω_i and ω_j are the statistical weights of the lower i and upper j levels, respectively. Similarly, the oscillator strength (f_{ij}) ...

(Dimensionless) and the line strength S (in atomic units, $1 \text{ a. u.} = 6.46\times10^{-36} \text{ cm}^2 \cdot \text{eus}^2$) are related by the following standard equations. For the electric dipole (E1) transitions,

$$A_{ji} = \frac{2.0261\times10^{18}}{\omega_j \lambda_{ij}^3} S^{E1} \text{ and } f_{ij} = \frac{303.75}{\omega_j \lambda_{ij}^3} S^{E1}$$

The wavelengths, transition probabilities, and weighted oscillator strengths, for E1transitions calculated using the (FAC) are reported in Table 2 and Table 3 for Ge XXIII. We present a comparison between our results and other theoretical transition probabilities for allowed E1 transitions for Ge XXIII in tables respectively. It shows that our results are in a good agreement with the other theoretical and experimental results [22] [23] . Table 2 shows a comparison between the FAC, MCDF, RCI, and other theoretical weighted oscillator strength [22] [23] values for some transitions among the levels of the 1s2 2s2 2p5nl, 1s2 2s1 2p6nl (n = 3, 4, 5, 6; l = s, p, d, f, g and h) configurations for Ge XXIII. Generally, there are no discrepancies between our FAC, COWAN, CIV3, MCDF, RCI, and experimental calculations and the agreement is within ≤20% for strong transitions.

3.3. Level Population

The reduced population densities are calculated for 241 finestructure levels arising from 1s2 2s2 2p6nl, 1s2 2s2 2p5 nl, 1s2 2s1 2p6 nl (n = 3, 4, 5, 6; l = s, p, d, f, g and h) configurations that emit radiation in the XUV and soft X-ray spectral regions. The calculations were performed by solving the coupled rate Equation (1) simultaneously using MATLAB R2013acomputer program. The present calculations for the reduced populations as a function of electron densities are plotted in Figure 1 to Figure 4 for levels (5, 15, 33, 37) of the configurations [(2p1\23s1\2)1,(2p1\23p1\2)0,(2s1\23p1\2)1,(2s1\23d5\2)2] at different plasma temperatures (650, 850, 1050, 1250, 1450,

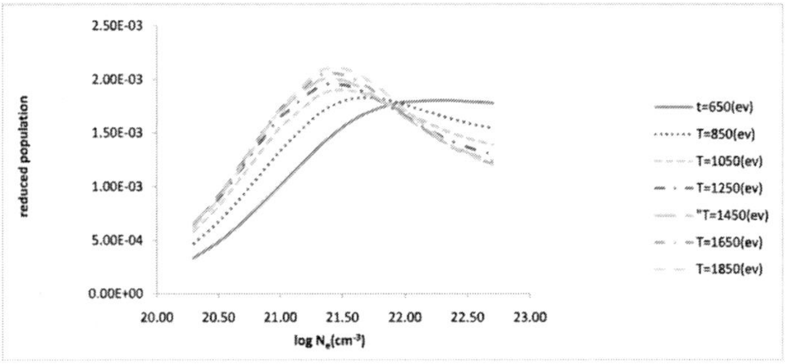

Figure 1. Reduced population of level (2p1\23s1\2)1 for Ge XXIII after electron collisional pumping as a function of the electron density at temperatures (650, 850, 1050, 1250, 1450, 1650, and 1850) eV.

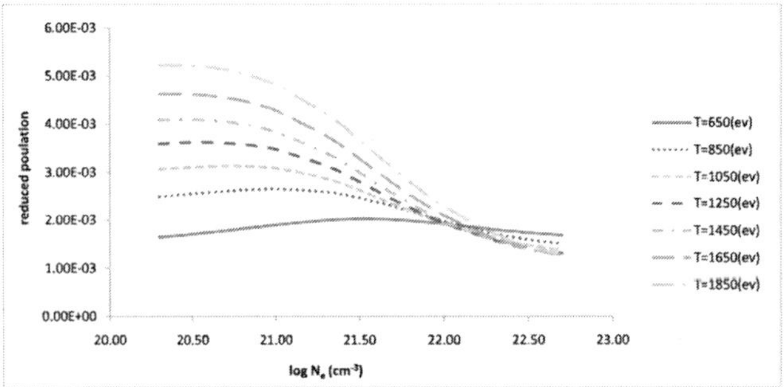

Figure 2. Reduced population of level (2p1\23p1\2)0 for Ge XXIII after electron collisional pumping as a function of the electron density at temperatures (650, 850, 1050, 1250, 1450, 1650, and 1850) eV.

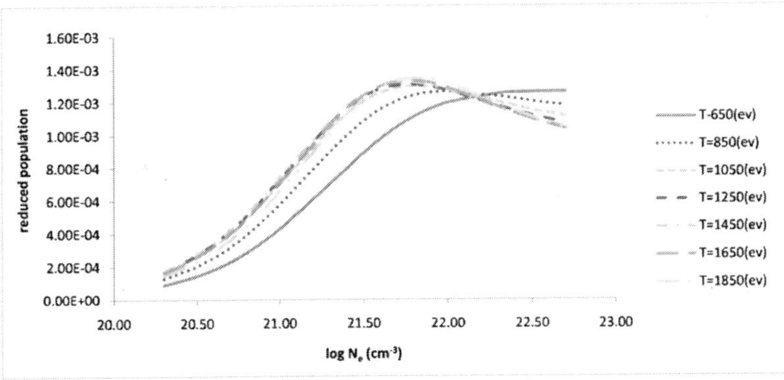

Figure 3. Reduced population of level (2s1\23p1\2)1 for Ge XXIII after electron collisional pumping as a function of the electron density at temperatures (650, 850, 1050, 1250, 1450, 1650, and 1850) eV.

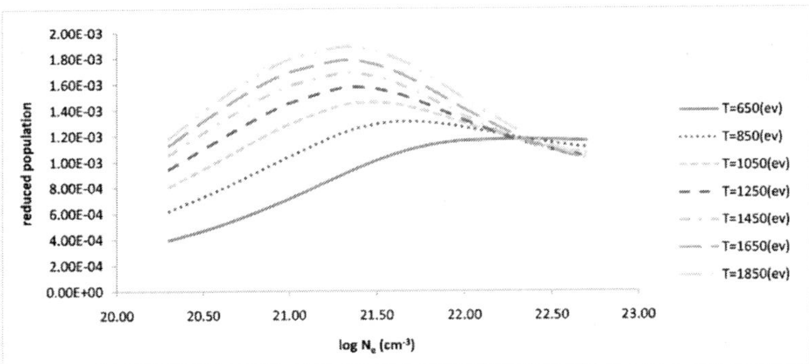

Figure 4. Reduced population of level (2s1\23d5\2)2 for Ge XXIII after electron collisional pumping as a function of the electron density at temperatures (650, 850, 1050, 1250, 1450, 1650, and 1850) eV.

1650, and 1850) eV for Ge XXIII ion. In the calculation we took into account spontaneous radiative decay rate and electron collisional processes between all levels under the study. The behavior of level populations of the various ions can be explained as follows: in general at low electron densities the reduced population density is proportional to the electron density, where excitation to an excited state is followed immediately by radiative decay, and collisional mixing of excited levels can be ignored. This result is in agreement with that of Feldman et al. [19] - [21] . At high electron densities (Ne > 5 × 1022), the radiative decay to all the levels will be negligible compared to collisional depopulations and all the level populations become independent of the electron density and are approximately equal (see Figures 1 to 4).

3.4. Radiative Lifetime

The lifetimes are determined almost entirely from the allowed and the strong inter combination transitions. The radiative lifetime τ_j of an excited atomic state j, is related to the atomic transition probability A_{ji} by:

$$\tau_j = \frac{1}{\sum_i A_{ji}} \tag{9}$$

where the sum is extended over all the lower states which can be reached from the upper state by radiative decay.

Table 4 contains the present results of radiative lifetime for the upper and lower laser levels for the Ge XXIII.

3.5. Inversion Factor

As we mentioned before, laser amplification will occur only if there is population inversion, or in other words, for positive inversion factor F > 0. However, large amplification, the gain exceeds all losses and ensures laser emission. In order to work in the XUV and X-ray spectral regions, we have selected transitions between any two levels producing photons with wavelengths between 12 and 52 nm. The electron density at which the population reaches corona equilibrium approximately equals to A/D, where A is the radiative decay rate and D is the collisional de-excitation rate [19] [21] . The population in version is largest where the electron collisions de-excitation rate for the upper level is comparable to the radiative decay rate for this level.

Table 4. Laser transition, wavelength,radiative life time of the upper and lower laser level in the possible laser transitions and maximum gian coefficient at temperatures (650,850,1050,1250,1450,1650, and 1850) eV in possible laser transition.

Transition	Configuration	λ (nm)	τ_u (sec)	τ_l (sec)	Gain (α) (cm^{-1}) T = 650 (ev)	T = 850 (ev)	T = 1050 (ev)	T = 1250 (ev)	T = 1450 (ev)	T = 1650 (ev)	T = 1850 (ev)
7<>3	$(2p_{3/2}3p_{1/2})_2$-----$(2p_{3/2}3s_{1/2})_1$	28.609	1.34E−10	3.50E−13	7.5414	11.642	12.694	15.1191	17.125	21.4424	27.3609
9<>3	$(2p_{1/2}3p_{1/2})_1$-----$(2p_{3/2}3s_{1/2})_1$	24.728	9.66E−11	3.50E−13	1.8221	3.3047	3.98416	5.1061	6.12157	7.73807	9.7982
10<>3	$(2p_{3/2}3p_{3/2})_2$-----$(2p_{3/2}3s_{1/2})_1$	23.189	7.00E−11	3.50E−13	6.9387	10.867	11.8141	14.2795	16.3407	20.1334	26.2206
15<>3	$(2p_{1/2}3p_{1/2})_0$-----$(2p_{3/2}3s_{1/2})_1$	12.741	2.31E−11	3.50E−13	6.2797	8.4190	8.36027	9.22122	10.2872	12.461	15.8628
14<>5	$(2p_{1/2}3p_{3/2})_2$-----$(2p_{1/2}3s_{1/2})_1$	23.584	8.07E−11	4.84E−13	4.3045	7.6260	8.65851	11.4741	13.9279	17.4553	21.512
15<>5	$(2p_{1/2}3p_{1/2})_0$-----$(2p_{1/2}3s_{1/2})_1$	18.730	2.31E−11	4.84E−13	23.865	29.814	28.6124	30.9059	33.8385	40.33	50.4729
21<>9	$(2p_{3/2}3d_{5/2})_2$-----$(2p_{1/2}3p_{3/2})_1$	20.192	4.96E−11	9.66E−11	1.1868	2.415	2.95213	3.98613	5.01531	6.66205	9.3533
35<>31	$(2s_{1/2}3d_{3/2})_2$-----$(2s_{1/2}3p_{1/2})_1$	19.565	4.72E−12	4.26E−13	0.7458	1.7933	2.7096	3.52883	4.33676	5.74496	7.30669
37<>31	$(2s_{1/2}3d_{5/2})_2$-----$(2s_{1/2}3p_{1/2})_1$	17.369	4.40E−12	4.26E−13	1.7522	2.6849	3.21201	3.9254	4.57372	5.37767	6.34003
37<>33	$(2s_{1/2}3d_{5/2})_2$-----$(2s_{1/2}3p_{3/2})_1$	19.990	4.40E−12	1.04E−13	27.958	41.249	37.1457	53.8662	61.1486	72.0261	87.8871
45<>39	$(2p_{3/2}4p_{3/2})_0$-----$(2p_{3/2}4s_{1/2})_1$	45.130	1.09E−12	5.01E−13	1.3883	1.9549	2.4693	3.56572	4.27755	4.85807	5.29178
123<>96	$(2s_{1/2}4d_{3/2})_2$-----$(2s_{1/2}4p_{1/2})_1$	51.643	4.96E−13	6.92E−14	1.787	4.3711	5.82193	7.66273	9.93204	13.4196	18.2148

where τ_u is the radiative life time of the upper laser level; τ_l is the radiative life time of the lower laser level; λ is the laser transition wavelength in (nm); α is the gain coefficient in cm^{-1}.

3.6. Gain Coefficient

As a result of population inversion there will be positive gain in laser medium. Equation (8) has been used to calculate gain coefficient for the Doppler broadening of various transitions in the Ge XXIII ion. Our results for the maximum gain coefficient in cm−1 for those transitions having appositive inversion factor F > 0 in the case of Ge XXIII ion at different temperatures are calculated (see Table 4) and plotted against electron density in Figure 5 and Figure 6. The figures show that the population inversions occur for several transitions in the Ge

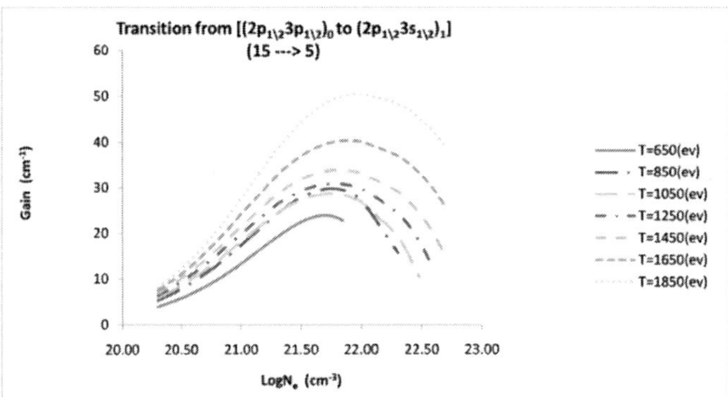

Figure 5. Gain coefficient against electron density (Ne) at different temperatures (650, 850, 1050, 1250, 1450, 1650, 1850) eV.

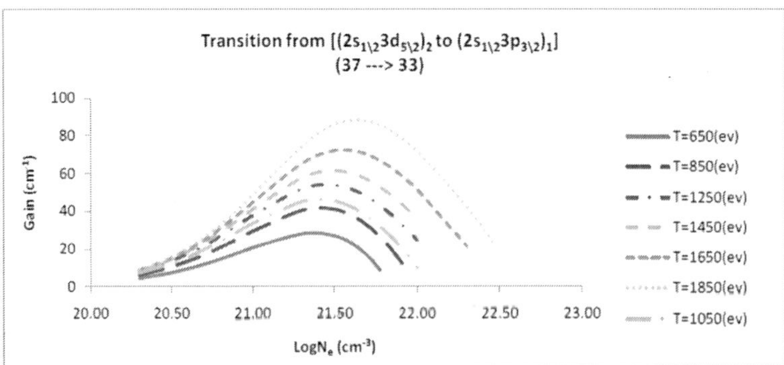

Figure 6. Gain coefficient against electron density (Ne) at different temperatures (650, 850, 1050, 1250, 1450, 1650, 1850) eV.

XXIII ion, however, the largest gain occurs for the Ge XXIII ion at (2s1\23d5\2)2-----(2s1\23p3\2)1 transition. These short wavelength laser transitions can be produced using plasmas created by optical lasers as the lasing medium. For Ge XXIII ion the rates for electron collisional excitation from the 1s2 2s2 2p6 ground state to the 1s22s12p63d configuration are greater than the rates for excitation from the ground state to the 1s22s12p63p state. For electron densities and electron temperatures that are typical of laboratory high-density plasma sources, such as laser produced plasmas, it is possible to create a quasi-stationary population inversion between the 2s1 2p63d and 2s1 2p63p states in Ge XXIII ion. Our calculations have shown that under favorable conditions large laser gains for this transition in the XUV and soft X-ray regions of the spectrum can be achieved in the Ge XXIII. The gain calculations were performed at electron temperatures equal to (650, 850, 1050, 1250, 1450, 1650, and 1850) eV at different electron densities. It is obvious that the gain increases with the temperature. We also find that some wavelengths for the lasing transitions in Table 4 are much closed to the values of it in the ref. [23] which give us the accurate wavelength calculated experimentally, although we Ignores some physical processes in the our rate equation.

4. CONCLUSIONS

The analysis of how the electron collisional pumping (ECP) is suitable for attaining population inversion and offering the potential for laser emission in the spectral region between 12 and 53 nm from the Ge XXIII ion.

This class of lasers can be achieved under the suitable conditions of pumping power as well as electron density. If the positive gains obtained previously for some transitions in the ions under studies (Ge XXIII ion) together with the calculated parameters can be achieved experimentally, then successful low-cost electron collisional pumping XUV and soft X-ray lasers can be developed for various applications. The results have suggested the laser transitions in the Ge XXIII plasma ion (see Table 4), as the most promising laser emission lines in the XUV and soft X-ray spectral regions.

REFERENCES

1. Gudzenko, L.I. and Shelepin, L.A. (1966) Amplification of Radiation in the Decay of a Maximally Ionized Plasma. Journal of Applied Mechanics and Technical Physics, 7, 82.
2. Maxon, S., Hagelstein, P., Reed, K. and Scofield, J. (1985) A Gas Puff Soft X-Ray Laser Target Design. Journal of Applied Physics, 57, 971.
3. Vinogradov, A.V., Sobelman, I. and Yukov, E. (1977) Population Inversion of Transitions in Neon-Like Ions. Soviet Journal of Quantum Electronics, 7, 32.
4. King, R.E., Pert, G.J., McCabe, S.P., Simms, P.A., MacPhee, A.G., Lewis, C.L.S., et al. (2001) Saturated X-Ray Lasers at 196 A and 73 A Pumped by Picosecond Travelling Wave Excitation. Physical Review A, 64, Article ID: 053810.
5. Rocca, J.J., Shlyaptsev, V., Tomasel, F.G., Cortázar, O.D., Hartshorn, D. and Chilla, J.L.A. (1994) Demonstration of a Discharge Pumped Table Top Soft X-Ray Laser. Physical Review Letters, 73, 2192.
6. Tomsel, F.G., Rocca, J.J., Shlyaptsev, V.N. and Macchietto, C.D. (1997) Lasing at 60.8 nm in Ne-Like Sulfur Ions in Ablated Material Excited by a Capillary Discharge. Physical Review A, 55, 1437.
7. Healy, S.P., Janulewicz, K.A. and Pert, G.J. (1997) Short Wavelength Lasing on Collisionally Pumped, Highly Excited 2s Hole States of Neon Like Ions in Performed Plasmas Irradiated with Picosecond Pulses. Optics Communications, 144, 24-30.
8. Simms, P.A., McCabe, S. and Pert, G.J. (1998) Comparison of Neon Like Germanium Inner Shell and Nickel Like Gadolinium at 62 A and 66 A by Picosecond Heating of a Performed Plasma. Optics Communications, 153, 164-171.

9. Norton, B.A. and Peacock, N.J. (1975) Population Inversion in Laser-Produced Plasmas by Pumping with Opacity- Broadened Lines. Journal of Physics B, 8, 989.
10. Bhagavatula, V.A. (1976) Soft X-Ray Population Inversion by Resonant Photoexcitation in Multicomponent Laser Plasmas. Journal of Applied Physics, 47, 4535.
11. Nilsen, J., Beiersdorfer, P., Elliott, S.R., Phillips, T.W., Bryunetkin, B.A., Dyakin, V.M., Pikuz, T.A., Faenov, A.Ya., Pikuz, S.A., von Goeler, S., Bitter, M., Loboda, P.A., Lykov, V.A. and Politov, V.Yu. (1994) article title. Physical Review A, Article ID: 502143.
12. Qi, N. and Krishnan, M. (1988) Enhancement of Ultraviolet Laser Plasma Emission Produced in a Strong Static Electric Field. Optics Communications, 67, 124-128.
13. Feldman, U., Bhatia, A.K. and Suckewer, S. (1983) Short Wavelength Laser Calculations for Electron Pumping in Neon-Like Krypton (Kr XXVII). Journal of Applied Physics, 54, 2188.
14. Feldman, U., Seely, J.F. and Bhatia, A.K. (1984) Scaling of Collisionally Pumped 3s-3p Lasers in the Neon Isoelectronic Sequence. Journal of Applied Physics, 56, 2475.
15. Chapline, G. and Wood, L. (1975) X-Ray Lasers. Physics Today, 28, 40. http://dx.doi.org/10.1063/1.3069004
16. Vinogradov, A.V. and Shlyaptsev, V.N. (1980) Calculations of Population Inversion Due to Transitions in Multiply Charged Neon-Like Ions in the 200-2000 Å Range. Soviet Journal of Quantum Electronics, 10, 754.
17. Goldstein, W.H., Oreg, J., Zigler, A., Bar-Shalom, A. and Klapisch, M. (1988) Gain Predictions for Nickel-Like Gadolinium from a 181-Level Multiconfigurational Distorted-Wave Collisional-Radiative Model. Physical Review A, 38, 1797.
18. Sobel’man, I.I. (1979) Introduction to the Theory of Atomic Spectra, International Series of Monographs in Natural Philosophy. Vol. 40, Pergamon Press, Oxford.
19. Hagelstein, P.L. (1986) Relativistic Distorted-Wave Results for Nickel-Like Gadolinium. Physical Review A, 34, 874.

20. Zhang, H.L., Samson, D.H. and Fontes, C.J. (1991) Relativistic Distorted-Wave Collision Strengths and Oscillator Strengths for the 33 Ni-Like Ions with 60 ≤ Z ≤ 92. Atomic Data and Nuclear Data Tables, 48, 91-163.
21. Feldman, U., Doschek, G.A., Seely, J.F. and Bhatia, A.K. (1985) Short Wavelength Laser Calculations for Electron Pumping in Be I and B I Isoelectronic Sequences (18 ≤ Z≤ 36). Journal of Applied Physics, 58, 2909.
22. Hibbert, A., Ledourneuf, M. and Mohan, M. (1993) Energies, Oscillators Strengths, and Life Times for Neon-Like Ions up to Kr XXVII. Atomic Data and Nuclear Data Tables, 53, 23-112.
23. Daido, H., Kodama, R., Murai, K., Yuan, G., Takagi, M., Kato, Y., Choi, I.W. and Nam, C.H. (1995) Significant Improvement in the Efficiency and Brightness of the J = 0-1 19.6-nm Line of the Germanium Laser by Use of Double-Pulse Pumping. Optics Letters, 20, 61-63.
24. NIST. http:///F:/NIST/NIST%20ASD%20Levels%20Output32.htm

Chapter 9

Growth and Characterization of Cu(In,Ga)Se2 Thin Films by Nanosecond and Femtosecond Pulsed Laser Deposition

Shih-Chen Chen[2], Dan-Hua Hsieh[1], Hsin Jiang[2], Yu-Kuang Liao[2], Fang-I Lai[4], Chyong-Hua Chen[1], Chih Wei Luo[2], Jenh-Yih Juang[2], Yu-Lun Chueh[3*], Kaung-Hsiung Wu[2*] and Hao-Chung Kuo[1*]

(1) Department of Photonics and Institute of Electro-Optical Engineering, National Chiao-Tung University
(2) Department of Electrophysics, National Chiao-Tung University
(3) Department of Materials Science and Engineering, National Tsing Hua University
(4) Department of Photonic Engineering, Yuan-Ze University

ABSTRACT

In this work, $CuIn_{1-x}Ga_xSe_2$ (CIGS) thin films were prepared by nanosecond (ns)- and femtosecond (fs)-pulsed laser deposition (PLD) processes. Different film growth mechanisms were discussed in perspective of the laser-produced plasmas and crystal structures. The fs-PLD has successfully improved the inherent flaws, $Cu_{2-x}Se$, and air voids ubiquitously observed in ns-PLD-derived CIGS thin films. Moreover, the prominent antireflection and excellent crystalline structures were obtained in the fs-PLD-derived CIGS thin films. The absorption spectra suggest the divergence in energy levels of radiative defects brought by the inhomogeneous distribution of elements in the fs-PLD CIGS, which has also been supported by comparing photoluminescence (PL) spectra of ns- and fs-PLD CIGS thin films at 15 K. Finally, the superior carrier transport properties in fs-PLD CIGS were confirmed by fs pump-probe spectroscopy and four-probe measurements. The present results indicate a promising way for preparing high-quality CIGS thin films via fs-PLD.

Keywords CIGS Pulsed laser deposition Femtosecond Photoluminescence Pump probe

BACKGROUND

$CuIn_{1-x}Ga_xSe_2$ (CIGS) has been extensively regarded as the most favorable absorber layer for thin film photovoltaic devices. CIGS possesses superior absorption characteristics due to its direct bandgap, which can be engineered by the partial substitution of indium by gallium atoms. Recently, the reported thin film CIGS-based solar cells have achieved the highest efficiency of 20.8% among all thin film solar cells at laboratory level[1]. The absorber layers for high-performance CIGS-based solar cells are usually prepared by vacuum processes (such as co-evaporation or sputtering). However, post-selenization and precise control of deposition parameters are required in both vacuum approaches[2, 3]. In contrast, pulsed laser deposition (PLD) is an alternative way that possesses the advantages of simple usage and good transfer of stoichiometry of target composition without post-selenization[4, 5]. All of these advantages are beneficial to obtain high-quality and reproducible CIGS thin films at low cost and are also suitable for investigating the underlying physical mechanisms that limit the efficiency.

The first PLD CIGS thin films were reported by Kusmartseva et al.; they investigated the effects of growth temperature and substrate material on the films[5]. Subsequently, extensive efforts have been devoted to study the correlations between the properties of PLD CIGS thin films and their growth parameters (such as composition of target, target-substrate distance, annealing time, laser repetition rate, and energy fluence)[6–9]. The laser sources used in the early-stage research were excimer lasers with pulse duration of tens of nanosecond (ns). Recently, Tsai et al. deposited CIGS thin films utilizing the femtosecond (fs) mode-locked Ti:sapphire laser rather than the excimer laser[10]. The annealing effects on the crystallization, microstructure, surface composition, and photoelectrical property of CIGS films were reported.

Comparisons of the growth mechanisms of PLD processes with different laser sources are crucial to understand the characteristics of high-quality CIGS thin films. However, the effect of laser sources on characteristics and growth mechanism of PLD CIGS thin films has not been investigated yet. In this study, the PLD CIGS thin films prepared by ns excimer laser and fs Ti:sapphire laser are compared to delineate the intimate correlations between the film growth mechanism and the associated plasma dynamics. Surface morphologies and grain structures were examined by scanning electron microscope (SEM). The crystalline structures and elemental distributions were analyzed by X-ray diffraction (XRD) spectrum and energy dispersive spectrum (EDS). The reflectance of the PLD CIGS films was measured by UV-visible-near infrared (NIR) spectrophotometer. In addition, defects and their effect on carrier dynamics were measured and investigated by photoluminescence (PL) and fs pump-probe spectroscopy.

METHODS

Growth of PLD CIGS thin films

The laser sources used in this experiment were KrF excimer laser (wavelength = 248 nm, pulse width = 20 ns, fluence approximately 10 J/cm^2, pulse repetition rate = 10 Hz) and Ti:sapphire mode-locked laser (wavelength = 800 nm, pulse width = 100 fs, fluence approximately 1 J/cm^2, pulse repetition rate = 5,000 Hz), which were for ns-PLD and fs-PLD growth, respectively. The CIGS thin films were deposited on soda-lime glass (SLG) substrates in a vacuum chamber. The background pressure was kept at 4×10^{-6} Torr. The distance between the target and substrate was 4 cm. The substrate temperature was monitored by a thermocouple attached to the substrate holder and was kept at optimal temperature of 500°C during the deposition processes.

Characterization

Powder XRD was conducted by a Bruker D2 PHASER X-ray spectrometer (Ettlingen, Germany) under irradiation of mono-chrome Cu-K_α (λ approximately 1.54 Å). The morphology, nanostructure, and elemental compositions of chalcopyrite films were obtained by field emission SEM (JSE-7001, JEOL, Tokyo, Japan) with attached accessory of EDS (INCA analysis system, Oxford Instruments, Oxfordshire, UK). Reflectance spectroscopy was acquired by the spectrophotometer (U-4100 UV-visible-NIR spectrophotometer, Hitachi, Tokyo, Japan). PL measurements were carried out using a 635-nm CW diode laser excitation source. The signal was dispersed by a monochromator and detected by an InGaAs photodiode (working range = 900 to 2,100 nm). A continuous helium cryostat was used for low-temperature PL measurements. For pump-probe measurements, a commercial Ti:sapphire laser system providing short pulses (approximately 30 fs) with repetition rate of 75 MHz and wavelength of 800 nm (hv = 1.55 eV) was used. The pump beam was focused at a diameter of about 50 μm with pump fluence ranging from 15.2 to 45.7 μJ/cm^2, while the probe fluence was fixed at approximately 1 μJ/cm^2 at spot diameter of 20 μm. The pump pulses were modulated at 2 KHz with a chopper. A mechanical delay stage was used to vary the time delay between the pump and probe pulses. The transient reflectivity change $\Delta R/R$ of the probe beam was measured as a function of the pump-probe delay time. The small reflected signals were detected and fed into a lock-in amplifier.

Results and discussion

Figure 1a,b shows the laser-produced plasmas (LPP) at the surface of the CIGS target by ns and fs laser, respectively. It exhibits substantial dissimilarities in LPPs that can be explained by the various laser-target interactions. For the ns-PLD process (Figure 1a), there is much residual heat, which is caused by the longer duration of laser pulse, as the pulse laser hits the target. The residual heat is due to the picosecond order of both the heat conduction time and ion energy transfer time, which is much faster than the pulse width of the excimer laser. It leads to the mixing of the melted CIGS (gray color) debris with the direct-transferred undesirable Cu_2Se secondary phases (yellow color) from the target as clusters were ejected along with the plasma in expansive directions. The effect of residual heat can spread to a wider range in the target, thus leading to an enlarged heat-affected zone (HAZ) (red region) that brings the plasma and debris with variation in energy and random transportation directions. This is why the expansive plasma was observed as shown in the inset of Figure 1a. Nonetheless, these large clusters can re-crystallize into a preferred orientation directed by the flow of the remaining residual energy of the laser pulses and the thermal energy from the heated substrate.

On the contrary, the highly localized interactions with target minimize the HAZ by the fs pulse laser. This is because the duration of laser pulse is shorter than the heat conduction time, so the residual energy can be eliminated. The main mechanism of producing plasma by fs pulse laser is coulomb explosion, a process that ionizes atoms in a solid-state target through an extremely intensive electric field, rather than conventional evaporation. With the absence of residual heat, concentrated plasma was generated by fs laser pulses (Figure 1b), which consists of the mixture of atoms and nanometer particles. These ingredients tend to construct the same crystal structure of the polycrystalline target when reaching the heated substrate due to the non-participation of residual heat as the re-crystallization energy. Similar properties of the fs pulse-induced laser plume were discussed by Verhoff et al[11].

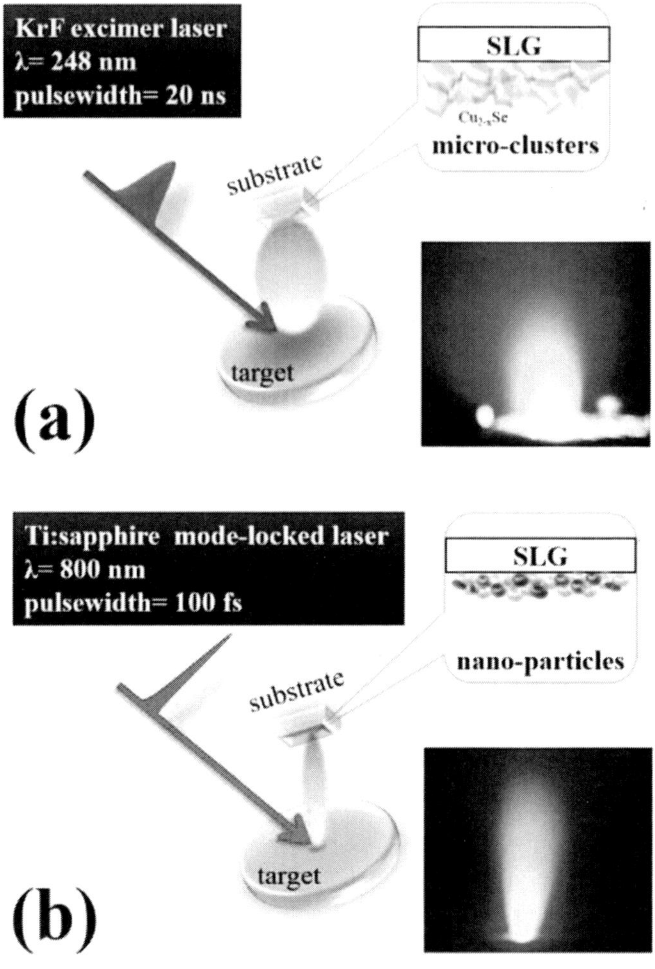

Figure 1 Schematic illustrations and photos of laser-produced plasmas on CIGS target. (a) ns-PLD and **(b)** fs-PLD.

Figure 2a,b shows the surface and grain morphologies of both ns-PLD and fs-PLD CIGS thin films. CIGS film deposited by the ns-PLD growth was found to have smooth surface and larger grain size, while much rougher surface with smaller grains was observed in films deposited by the fs-PLD growth. Figure 2c shows the side-view SEM image of the ns-PLD CIGS thin film, in which the grain boundaries (GBs) can be clearly observed. In contrast, the GBs of the fs-PLD CIGS thin film are barely seen as shown in Figure 2d, which indicates a more compact structure as expected. As shown in Figure 2a, there are a lot of micro-clusters generated due to the residual heat generated by ns laser pulses. It has also been found that the secondary phases ($Cu_{2-x}Se$) with Cu/In/Ga/Se = 62.92:1.42:0.82:34.84 characterized by EDS tend to segregate on the surface and appear as large droplets indicated by the white arrow shown in Figure 2a[9]. However, it is evident from Figure 2b that the segregation of secondary phases is significantly reduced in films obtained by fs-PLD[11]. Moreover, air voids occurring at grain boundaries (marked by the white arrow in the inset of Figure 1a) were observed in films deposited by the ns-PLD. The formation of air voids between grains is most likely due to the stack of the larger clusters and debris. It is worthy to note that both of the abovementioned microstructure features exhibited in films deposited by the ns-PLD can lead to substantial current leakage in devices. Such detrimental disadvantages, nevertheless, can be successfully removed with a concentrated and oriented plume consisting of atoms and nanometer-cluster mixtures resulting from the localized strong electric field ionization on the target by using the fs pulses[12]. In addition, ingredients of the nanometer-cluster mixture evidently resulted in a much more compact CIGS films (Figure 2b). Consequently, the inherent nanostructure uniformly distributed on the surface of fs-PLD-derived CIGS film is observed instead of the micrometer-sized droplets of the secondary phases.

The XRD patterns of the CIGS target and the two CIGS thin films are presented in Figure 3a. In the pattern of the CIGS target, the main peaks are broadened and degenerated to the peaks of binary crystals of $Cu_{2-x}Se_x$, which is commonly found in the hot-pressed CIGS pellet. In contrast, the homogeneous phase and remarkable crystallinity can be found in the two CIGS thin films. The polycrystalline feature with the chalcopyrite structure in the CIGS target is directly transferred to the CIGS films obtained by both ns- and fs-PLD processes. In addition, the full-width at half maximum (FWHM) of (112) diffraction peaks are 0.18° and 0.14° in ns-PLD and fs-PLD CIGS thin films, respectively. The smaller FWHM is indicative of larger grain size and better crystallinity in the fs-

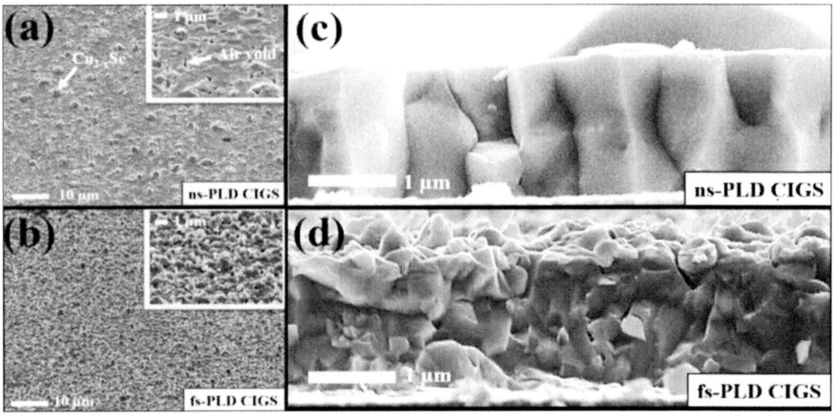

Figure 2. SEM images of ns-PLD CIGS and fs-PLD CIGS**.** Top-view SEM images of **(a)** ns-PLD CIGS and **(b)** fs-PLD CIGS. Side-view SEM images of (**c**) ns-PLD CIGS and (**d**) fs-PLD CIGS.

PLD CIGS. Furthermore, the existence of the (220)-oriented peak, which is beneficial for reducing the surface recombination of the CIGS absorber layer due to higher work function, is largely preserved only in films grown by the fs-PLD[13]. Preliminary studies have also shown that the relaxed structure usually accompanies with the broadened peak of (112) orientation, which is associated with high degree of structural disorder[14]. The high degree of structural disorder can be successfully suppressed for the fs-PLD CIGS thin film because of the well-crystalline characteristics confirmed by XRD spectra. The analyses of elemental composition ratios of CIG ([Cu]/[In] + [Ga]) and SCIG ([Se]/[Cu] + [In] + [Ga]) were carried out using the EDS measurements as shown in Figure 3b,c, respectively, where we randomly selected eight points from both PLD films for statistical analysis. It is observed that the ns-PLD CIGS film has more homogenous elemental distribution and is most likely due to the (112) dominant phase. Furthermore, compositions of copper and selenium of the ns-PLD CIGS film are averagely higher than that of the fs-PLD CIGS film. Other studies have reported the existence of more selenium deficiencies in PLD CIGS films[15]. This non-stoichiometry is more significant in the fs-PLD CIGS.

These results could be related to the high vapor pressure of selenium. When the target is under the fs laser irradiation, the atom and nanoparticle mixture is evaporated by ultrashort pulses. During the flight of the mixture to the substrate, 're-evaporation' of the nanoparticles happens and selectively decreases the elements in the mixture due to the insufficient energy that maintains the flight of the mixture to the substrate. The results agree with the fact that the pulse energy of the fs laser is much smaller than that of the ns laser (the pulse energy is 0.2 and 400 mJ for fs-PLD and ns-PLD, respectively). Re-evaporation can be significantly more effective in the mixture obtained by the fs laser pulses, which is of atomic and nanoparticle scale[14]. On the other hand, the secondary phase ($Cu_{2-x}Se$) clusters were 'ablated' from the target in the ns-PLD at its pristine phase (therefore, less re-evaporation can cause element loss). Moreover, the binary crystals also give rise to higher concentrations of copper and selenium in the thin film.

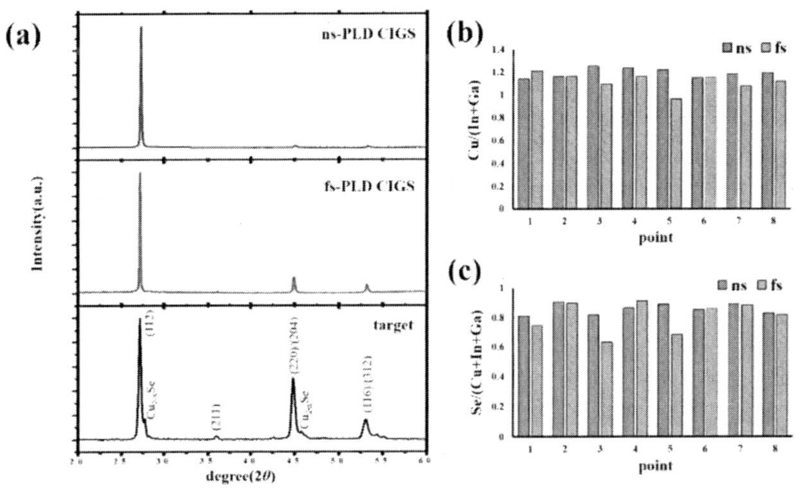

Figure 3. Material characterizations of target and both PLD films. **(a)** XRD spectra, **(b)** CIG ratio, and **(c)** SCIG ratio for both PLD films.

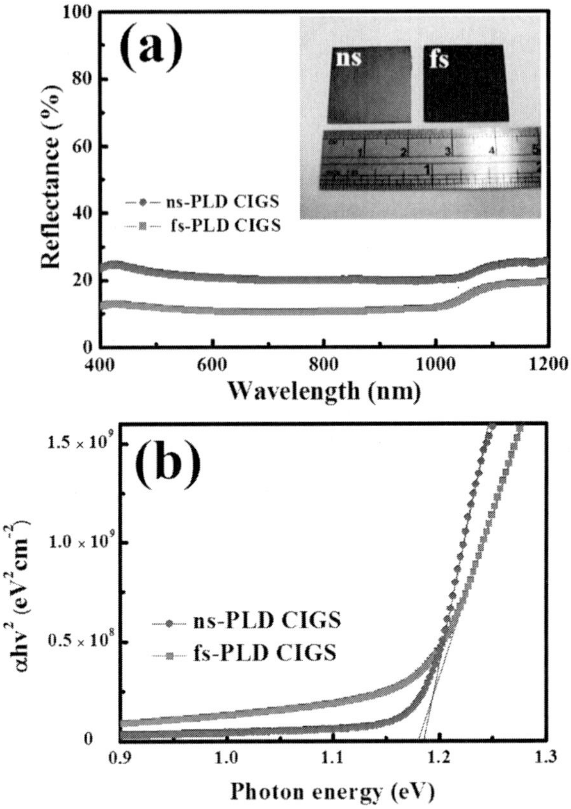

Figure 4. Reflectance (a) and absorption (b) spectra of ns- and fs-PLD CIGS thin films. The inset in (a) shows the photo of the two CIGS thin films.

The reflectance of the PLD CIGS thin films were measured as shown in Figure 4a. Obviously, the reduced reflectance is achieved in the fs-PLD CIGS film, as compared with that of the ns-PLD film. The reduced reflectance of the fs-PLD CIGS film can be attributed to the inherent nanostructure on the surface during the PLD growth process. The significant contrast in color also reveals the anti-reflection effect of the fs-PLD CIGS thin film, as shown in the inset of Figure 4a. It is a prominent property compared to the nanostructured CIGS film prepared by an extra etching process[16]. In addition, the ns- and fs-PLD

CIGS thin films have a similar bandgap value of approximately 1.2 eV extracted from absorption spectra, as shown in Figure 4b. The value is well consistent with the bandgap of the target with elemental compositions of Cu/In/Ga/Se = 1:0.7:0.3:2, respectively, revealing that the variation in elementary compositions in the fs-PLD CIGS (Figure 3b,c) is localized, while the global composition of the film still remained unchanged with the same composition as that of the target. Furthermore, fs-PLD CIGS shows a longer absorption tail due to the more diverged sub-band gap energy levels of radiative defects, which is most likely resulted from the local inhomogeneous distributions of elements.

Many studies have suggested that the defects of CIGS thin films are crucial to the performance of their device performances. PL is a powerful tool to shed light on defects arising from the deviation of stoichiometry[17]. Figure 5a shows the PL spectra of fs- and ns-PLD CIGS thin films at 15 K and room temperature (see the inset) without normalization, in which PL peaks at 1.2 eV for ns-PLD CIGS agrees well with the bandgap value obtained from the absorption spectrum (Figure 4b). Hence, we assign this peak as band-to-band transition, and other PL emission peaks with energy lower than 1.2 eV are assigned to different radiative defect-related transitions. At 15 K, where transitions between the defect levels are the dominant processes for CIGS, the intensity of the two PL spectra is comparable, suggesting that the defect type and concentration in the two samples are similar. By comparison, it can be seen that individual PL emission peaks can only be resolved from the PL spectrum of the ns-PLD CIGS, while no discrete PL emission peaks can be observed from that of the fs-PLD CIGS thin film. This could be due to the fluctuations of defect energy levels in the fs-PLD CIGS thin film, which broadens the FWHM of the PL emission peaks associated with all radiative defect-related transitions. The increased overlapping of the PL emission peaks, in turn, results in the unresolvable spectrum. Such fluctuations in radiative defect energy levels have also been observed in the absorption spectrum of the fs-PLD CIGS thin film shown in Figure 4b. The absorption spectrum of the fs-PLD CIGS shows a tail at

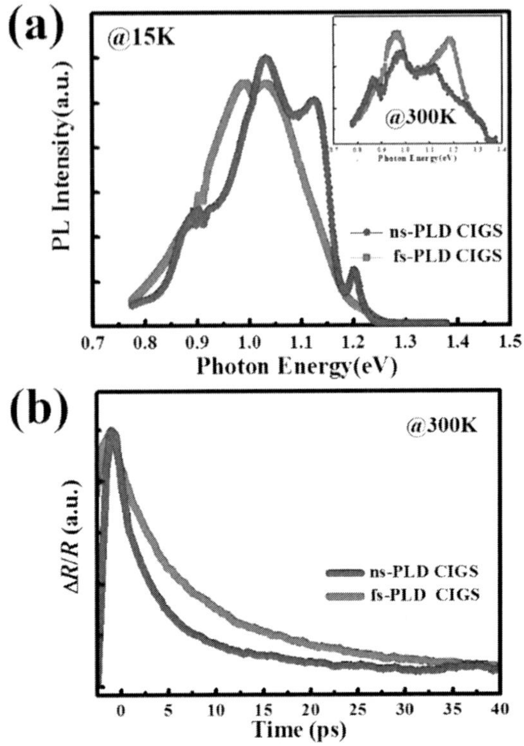

Figure 5. PL spectra (a) and fs pump-probe spectra (b) for ns-PLD (blue) and fs-PLD (red) CIGS thin films.

energies below its bandgap energy of 1.2 eV, indicating that energy level fluctuations related to the sub-band gap radiative defects might have been responsible for the photon absorptions observed at energies smaller than the intrinsic CIGS bandgap energy. The absorption tail can also be observed in the absorption spectrum of the ns-PLD CIGS thin film. Yet, the tail is much less significant for the ns-PLD CIGS film, presumably due to the fact that the individual radiative defect energy levels in ns-PLD CIGS film are more concentrated and less fluctuating. The discreteness of the PL emission peaks seen in the PL spectrum of the ns-PLD CIGS films evidently lends strong support to the above conjecture. At room temperature, the ns-PLD CIGS film shows a weaker PL intensity than that of the fs-PLD CIGS, which is due to the higher concentration of non-radiative recombination centers induced by surface state between CIGS/$Cu_{2-x}Se$ and CIGS/void interfaces. In addition, the stronger PL intensity of the fs-PLD CIGS can correspond to the existence of the (220)-oriented peak whose higher work function is beneficial for reducing the surface recombination. The results indicate that the fs-PLD CIGS film is much more promising for device performance compared to the ns-PLD CIGS film.

The defects in the CIGS thin films can also affect the carrier dynamics, hence their device performance. To this respect, carrier dynamics in CIGS thin films obtained by different PLD processes were investigated by fs pump-probe spectroscopy, which is a technique ubiquitously adopted to delineate the non-equilibrium carrier dynamics in semiconductors[18, 19]. Figure 5b shows the reflectivity transient in both films with a pumping power of 30.4 μJ/cm^2 at room temperature. It is apparent from Figure 5b that the carrier lifetime is much longer in the fs-PLD CIGS film. The defect-related non-radiative recombination lifetime (τ_n) can be derived from the results obtained by using different pumping fluences. It showed that the τ_n of ns- and fs-PLD CIGS films are 20 and 30 ps, respectively, revealing that the Shockley-Read-Hall (SRH) mechanism is more dominant in the ns-PLD CIGS at room temperature because of the existence of CIGS/$Cu_{2-x}Se$ and CIGS/void interfaces. On the other hand, the longer lifetime in the fs-PLD CIGS suggests less SRH recombination that is consistent with the existence of the (220) orientation. Finally, we examined the electrical properties by van der Pauw four-probe measurements. The resistivity values of ns- and fs-PLD CIGS thin films were approximately 66.0 Ω cm and approximately 0.1 Ω cm, respectively. The higher resistivity of the ns-PLD CIGS thin films can be attributed to the higher concentration of non-radiative recombination center verified by PL and pump-probe measurements. The superior carrier transport properties exhibited in the fs-PLD CIGS film again could be attributed to the substantial improvements realized in suppressing the formation of $Cu_{2-x}Se$ secondary phase and air voids by the fs-PLD process.

CONCLUSIONS

In summary, comparative investigations on the structure-property relations of CIGS thin films prepared by ns- and fs-PLD processes were conducted. The residual heat remaining on the target due to pulse duration difference was found to result in drastically different appearance of the laser-produced plasmas; hence, it led to vastly different film growth mechanisms and eventual film microstructures. The CIGS thin film prepared by fs-PLD, as compared to that obtained by the ns-PLD process, evidently exhibits much better film quality and superior carrier transport properties, primarily due to the removal of $Cu_{2-x}Se$ and air voids. In addition, the fs-PLD CIGS thin film also exhibits significantly better antireflection characteristic over a wavelength range of 400 to 1,200 nm. The absorption spectra suggest the divergence in energy levels of radiative defects brought by the inhomogeneous distribution of elements in fs-PLD CIGS. Such inference is strongly supported by comparing the PL spectra between the ns- and fs-PLD CIGS thin films at 15 K. Room temperature PL spectra of ns- and fs-PLD CIGS thin films suggest that in the ns-PLD CIGS films, there might exist more surface states at CIGS/$Cu_{2-x}Se$ and CIGS/void interfaces, which may act as the non-radiative recombination centers. Finally, fs pump-probe spectroscopy and four-probe measurements reveal that the fs-PLD CIGS films have a much longer carrier lifetime and significantly lower resistivity, both are beneficial for photovoltaic applications. The present results convincingly indicate that the fs-PLD process is a very promising method for preparing high-quality CIGS thin films.

ABBREVIATIONS

CIGS: $CuIn_{1-x}Ga_xSe_2$

EDS: energy dispersive spectrum

fs: femtosecond

*FWHM:*full-width at half maximum

GBs: grain boundaries

HAZ: heat-affected zone

LPP: laser-produced plasmas

ns: nanosecond

PL: photoluminescence

PLD: pulsed laser deposition

SEM: scanning electron microscope

SLG: soda-lime glass

XRD: X-ray diffraction.

ACKNOWLEDGEMENTS

The research was supported by the Ministry of Science and Technology through Grant Nos. 102-2112-M-009-006-MY3, 101-2112-M-007-015-MY3, 101-2218-E-007-009-MY3, and 102-2633-M-007-002, and the National Tsing Hua University through Grant No. 102N2022E1. YLC greatly appreciates the use of facility at CNMM, the National Tsing Hua University through Grant No. 102N2744E1.

Authors' contributions

SCC, DHH, and HJ carried out the thin film growth, XRD, SEM, and pump-probe analysis, and drafted the manuscript. YKL and FIL carried out the PL analysis. CHC participated in the design of the study. YLC, CWL, JYJ, KHW, and HCK conceived the study and organized the final version of the paper. All authors read and approved the final manuscript.

REFERENCES

1. Jackson P, Hariskos D, Wuerz R, Wischmann W, Powalla M: Compositional investigation of potassium doped Cu(In, Ga)Se_2 solar cells with efficiencies up to 20.8%. Phys Status Solidi 2014, 8: 219–222.

2. Hanket GM, Shafarman WN, McCandless BE, Birkmire RW: Incongruent reaction of Cu–(InGa) intermetallic precursors in H_2Se and H_2S. J Appl Phys 2007, 102(7):4074922.

3. Alberts V, Titus J, Birkmire RW: Material and device properties of single-phase Cu(In, Ga)(Se, S)2 alloy prepared by selenizationy/sulfurization of metallic alloys. Thin Solid Films 2004, 451–452: 207–211.

4. Dijkkamp D, Venkatesan T, Wu XD, Shaheen SA, Jisrawi N, Min-Lee YH, Mclean WL, Croft M: Preparation of Y-Ba-Cu oxide superconductor thin films using pulsed laser evaporation from high T_c bulk material. Appl Phys Lett 1987, 51: 619–621. 10.1063/1.98366

5. Levoska J, Leppavuori S, Wang F, Kusmartseva O: Pulsed laser ablation deposition of $CuInSe_2$ and $CuIn_{1-x}Ga_xSe_2$ thin films. Phys Scripta 1994, T54: 244–247.

6. Kindyak AS, Kindyak VV, Gremenok VF: Energy-gap variations in thin laser-deposited Cu (In, Ga)Se2 films. Mater Lett 1996, 28: 273–275. 10.1016/0167-577X(96)00065-1

7. Yoshida A, Tanahashi N, Tanaka T, Demizu Y, Yamamoto Y, Yamaguchi T: Preparation of $CuInSe_2$ thin films with large grain by excimer laser ablation. Sol Energy Mater Sol Cells 1998, 50: 7–12. 10.1016/S0927-0248(97)00093-7

8. Victora P, Nagarajub J, Krupanidhia SB: Pulsed excimer laser ablated copper indium diselenide thin films. Solid State Commun 2000, 116: 649–653. 10.1016/S0038-1098(00)00409-9

9. Jo YH, Mohanty BC, Cho YS: Enhanced electrical properties of pulsed laser-deposited $CuIn_{0.7}Ga_{0.3}Se_2$ thin films via processing control. Sol Energy 2010, 84: 2213–2218. 10.1016/j.solener.2010.08.005

10. Tsai MG, Tung HT, Chen IG, Chen CC, Wu YF, Qi X, Hwu Y, Lin CY, Wu PH, Cheng CW: Annealing effect on the properties of $Cu(In_{0.7}Ga_{0.3})Se_2$ thin films grown by femtosecond pulsed laser deposition. J Am Ceram Soc 2013, 96: 2419–2423. 10.1111/jace.12422

11. Verhoff B, Harilal SS, Freeman JR, Diwakar PK, Hassanein A: Dynamics of femto- and nanosecond laser ablation plumes investigated using optical emission spectroscopy. J Appl Phys 2012, 112: 093303. 10.1063/1.4764060

12. Balling P, Schou J: Femtosecond-laser ablation dynamics of dielectrics: basics and applications for thin films. Rep Prog Phys 2013, 76: 036502. 10.1088/0034-4885/76/3/036502

13. Ahmed E, Hill AE, Pilkington RD, Tomlinson RD, Leppavuori J, Levoska J, Kusmartseva O, Ahmed W, Afzal A: Deposition and characterization of copper indium gallium diselenide films by laser ablation and flash evaporation for use in solar cells. J Mater Sci 1997, 32: 5611–5613. 10.1023/A:1018684726577

14. Teghil R, D'Alessio L, De Bonis A, Galasso A, Ibris N, Salvi AM, Santagata A, Villani P: Nanoparticles and thin film formation in ultrashort pulsed laser deposition of vanadium oxide. J Phys Chem 2009, A113: 14969–14974.

15. Chaisitsak S, Yamada A, Konagai M: Preferred orientation control of $Cu(In_{1-x}Ga_x)Se_2$ ($x \approx 0.28$) thin films and its influence on solar cell characteristics. Jpn J Appl Phys 2002, 41: 507–513. 10.1143/JJAP.41.507

16. Liu CH, Chen CH, Chen SY, Yen YT, Kuo WC, Liao YK, Juang JY, Kuo HC, Lai CH, Chen LJ, Chueh YL: Large scale single-crystal

Cu(In, Ga)Se2 nanotip arrays for high efficiency solar cell. Nano Lett 2011, 11(10):4443–4448. 10.1021/nl202673k

17. Siebentritt S, Gütay L, Regesch D, Aida Y, Deprédurand V: Why do we make Cu(In, Ga)Se2 solar cells non-stoichiometric? Sol Energy Mater Sol Cells 2013, 119: 18–25.

18. Chen SC, Liao YK, Chen HJ, Chen CH, Lai CH, Chueh YL, Kuo HC, Wu KH, Juang JY, Cheng SJ, Hsieh YP, Kobayashi T: Ultrafast carrier dynamics in Cu(In, Ga)Se2 thin films probed by femtosecond pump-probe spectroscopy. Opt Express 2012, 20(12):12675–12681. 10.1364/OE.20.012675

19. Tisdale WA, Williams KJ, Timp BA, Norris DJ, Aydil ES, Zhu XY: Hot-electron transfer from semiconductor nanocrystals. Science 2010, 328: 1543–1547. 10.1126/science.1185509

Chapter 10

Comparison on Morphological and Optical Properties of TiO2 Thin Films Grown by Single-Pulse and Multi-Pulse Laser Ablation

Yonic Peñaloza-Mendoza, Luis Ponce-Cabrera

Instituto Politecnico Nacional, Cicata Altamira, Altamira, México

ABSTRACT

TiO_2 thin films were prepared on glass substrates using the PLD (Pulsed Laser Deposition) technique. In order to carry out the ablation process, a Nd:YAG laser was used emitting in 1064 nm wavelength at 10 Hz repetition rate, set up for operating in both single-pulse and multi-pulse regimes. A comparison of the deposition rate, the optical and morphological properties of the layers obtained from both ablation regimes was made, which showed that the multi-pulsed ablation produced layers with a higher surface quality and better optical properties.

Keywords: Pulsed Laser Deposition, Single-Pulse Ablation, Multi-Pulse Ablation, TiO_2 Thin Film

1. INTRODUCTION

Titanium dioxide (TiO_2) has become an extremely valuable metallic oxide for many technical applications in different uses due to its unique electrical and optical properties and the fact that it is not toxic and has a low cost [1] . TiO_2 thin films have a wide spectrum of applications in photocatalysis, photovoltaic and optical systems, especially in multilayer structures that can be anti-reflecting or have a high reflectance level at a specific wavelength [2] .

Previously, many techniques as sol-gel [2] , spray pyrolysis [3] , sputtering [4] and chemical vapor deposition [5] , have been applied to prepare TiO_2 thin films. A technique that has been increasingly applied for obtaining thin layers is the so-called Pulse Layer Deposition (PLD) [6] - [8] . Due to its versatility, this technique has extended to include the deposition of a great number of materials, ranging from pure elements to multicomponent compounds [9] [10] . Compared to other physical deposition techniques, PLD has the advantage of easily and amply controlling the physical, chemical and structural properties of many materials [11] , as well as in many cases the capacity to preserve the stoichiometry.

An important characteristic of PLD is the use of a pulsed laser as the source of excitation for ablating the target material, which under high vacuum conditions moves to the substrate, forming the layer with the desired thickness, ultimately proportional to the flow and amount of pulses.

Recently, Nd:YAG lasers with Q-Switch device based on saturable absorbers have been used for the purpose of exciting samples for elementary analysis by Laser Induced Breakdown Spectroscopy (LIBS). In this regime, the Q-Switch consists of a Chrome-doped YAG crystal (Cr^{4+}:YAG), which is clarified during the pumping process, emitting one or several pulses in the course of a one lamp's excitation pulse [12] [13] . Figure 1 shows the temporary structure of the pulses for both types of excitation, which we will call single-pulse and multi-pulse, respectively. Jedlinszki and Galbács show in [14] that in principle it is possible to reach higher levels of plasma ionization when the laser emits multi-pulse regime. This can be a very interesting point in the PLD technique, as higher plasma ionization levels, with the presence of very energy-rich species, may contribute to the growth of the film by transferring energy to the process.

This paper describes the use of the PLD technique for producing TiO_2 thin films with the two ablation regimes: single-pulse and multi-pulse, and the optical and morphological properties, as well as the deposition rates are evaluated for the purpose of establishing the potential of the multi-pulse regime as an excitation source in PLD.

2. EXPERIMENTAL PROCEDURE

A Nd:YAG laser with 1064 nm wavelength and 10 Hz repetition rate was used for the ablation process; it was set up with saturable Cr^{4+}:YAG absorbers for operating in single-pulse and multi-pulse operation regimes. In the multi-pulse regime, each laser shot is made up of three micro-pulses with an interpulse separation of 45 μs, where each micro-pulse has an average duration of 60 ns, as shown in Figure 1. The separation is 100 ms between each shot.

For the purpose of obtaining the single-pulse emission, a second Cr^{4+}:YAG Q-Switch was added to the optical cavity, which had the same dimensions and transmittance. This second Q-Switch increases the loss in the optical cavity, thus rendering the resupply process less efficient [15] , reducing the emission (previously multi-pulse) to just one laser pulse. Figure 2 shows the temporary structure of micro-pulse for single-pulse regimes, we observe that the pulse width was 40 ns. When compared to the multi-pulse regime, the reduction of the pulse width arises because when the second Q-Switch is incorporated, the total dimension of the saturable absorber increases, allowing a higher level of population inversion to be reached.

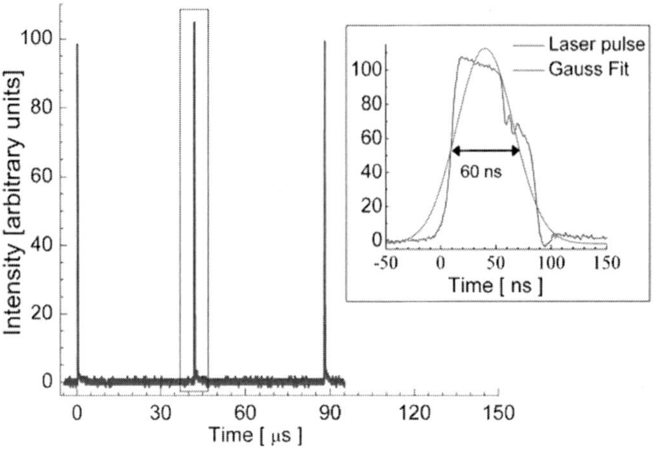

Figure 1. Temporary pulse profile and duration in the multi-pulsed ablation regime.

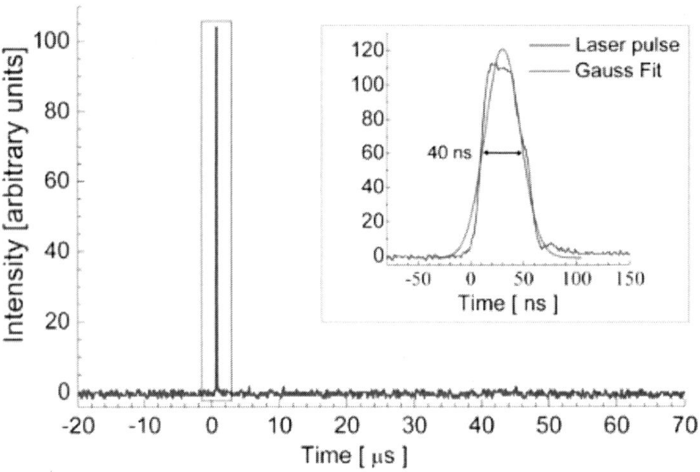

Figure 2. Temporary pulse profile and duration in the single-pulsed ablation regime.

The total energy of the pulse was set at 100 mJ for both regimes, obtaining 3.3×10^7 W·cm^{-2} and 5×10^7 W·cm^{-2} of irradiance, for multi-pulses and single-pulses respectively. The depositions were made on glass substrates in a high vacuum atmosphere, 2 - 5×10^{-5} Torr, at a constant flow and temperature, 2 J·cm^{-2} and 130°C. The target-substrate distance was set at 5 cm.

Depositions were made at different pulse quantities ranging from 3000 to 1200 pulses for the purpose of obtaining different layer thicknesses. The first characterization stage consisted of evaluating the morphology of the film by Atomic Force Microscopy (AFM), and also the roughness and the thickness. In a second stage an evaluation was made of the optical characteristics, comparing the absorbance of the films which had a similar thickness in the two regimes.

3. RESULTS AND DISCUSSION

The first criterion of the comparison was based on the morphological characteristics: roughness and thickness. In order to measure the thickness, small areas were removed after the deposition process was completed; this was done using laser pulses with the necessary irradiance for producing the complete ablation of the film on the area of the spot, without damage the substrate. This created a crater whose wall represents a reference step for thickness measuring, Figure 3, which was done with an atomic force microscope (TT-AFM Workshop) in contact mode.

Figure 4 shows the thickness as a function of the number of shots employed to make the thin films. Once the thickness measurements were available the deposition rates were calculated at 0.1 ± 0.01 Å/pulse and 0.04 ± 0.006 Å/pulse, for the single-pulse and multi-pulse regimes respectively.

The difference in the deposition rates may be because, taking into account the energy and vacuum conditions used, the separation between the micro-pulses in a multi-pulse shot is similar to the average life-time of the plasma [16] . The ablation process starts with the incidence of the first pulse on the target. When the second pulse is emitted, 45 μs later, the plasma is still in the extinction process, thus generating a pulse-plasma interaction. During this interaction, fraction of the energy of the pulse can be absorbed by the species still present in the plasma. Consequently, the energy impacting on the target is less and, therefore, the amount of extracted material is also less. This pulse-plasma interaction is unlikely in single-pulse regimen, because the separation between one pulse and next is 100 ms (10 Hz), a time largest than plasma life time (the range of μs).

Figure 5 reveals that for growths done with a lower number of shots (up to 6000 shots), the roughness level is similar in both regimes. On the other hand, for thicker layers (9000 and 12,000 shots), the roughness increased with the single-pulsed ablation.

Based on the pulse-plasma interaction described above, the species that absorbed part of the energy of the 2nd and 3rd pulse of the multi-pulse shot will increase their kinetic energy. This energy increase represents a longer diffusion time for the species because they are absorbed by the substrate surface. As the diffusion time increases, the probability that the atoms/molecules reach points of maximum absorption energy also increases, thus favoring an orderly growth of the nanolayers [6] [17] [18] . This process is evident in Figure 5, which shows that the roughness of the films grown with micro-pulses reveals very small variations when the deposition time increases, 0.5 Å/min, as compared to layers grown with multi-pulses, whose roughness grows 2.3 Å/min.

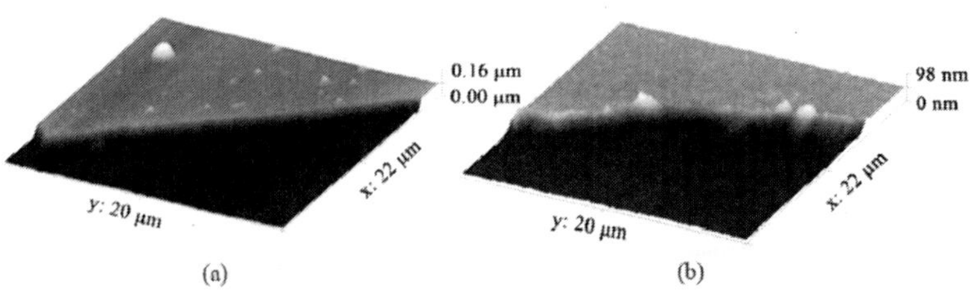

Figure 3. AFM images of TiO_2 thin films surface and reference step, made with: (a) 6000 shots in single-pulsed regime; (b) 12,000 shots in multi-pulsed regime.

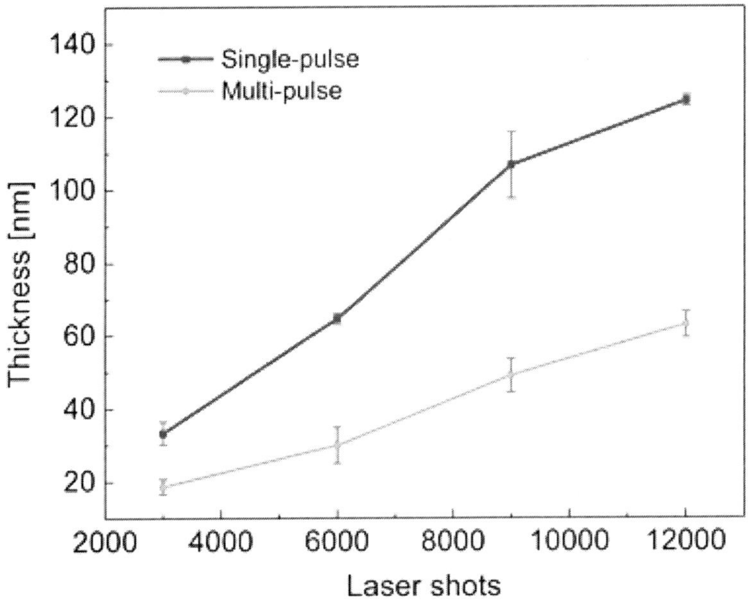

Figure 4. Thickness as a function of the number of shots of TiO_2 thin films made by single- pulse and multi-pulse ablation regimes.

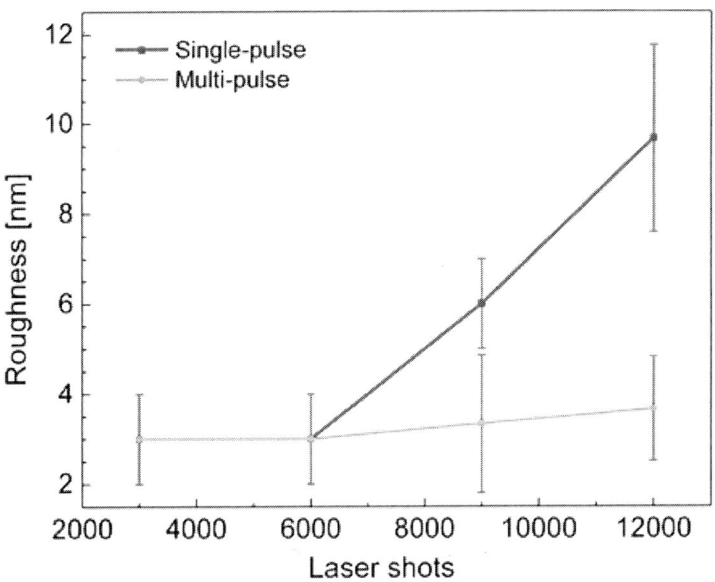

Figure 5. Roughness as a function of the number of shots of TiO_2 thin films made by single- pulse and multi-pulse ablation regimes.

The increase in the roughness with the single-pulse regime can be the result of two factors: first of all, as the deposition rate increases, the surface diffusion of the atoms decreases, and this could generate flaws on the surface of the nanolayers. The second factor is the presence of particulates or splashing, which as Figure 3 shows, is bigger in the films grown with multi-pulses. The presence of these micrometric particles in the initial stages of growth of the nanolayers may bring about an accumulative effect of defects on the final surface of the film, thus causing an accelerated increase in the roughness. In addition, it was suggested that the film roughness is influenced by strain induced in the film by a lattice mismatch with the substrate [19] , this affect the roughness both ablation regimens.

Figure 6 shows images obtained by scanning electron microscopy (SEM) of the films grown with 6000 and 12,000 shots, with a single-pulse and multi-pulse regime, respectively. The films are similar in thickness, 65 nm

and 63 nm, and have the same roughness, 3 nm. The thin films shown splashing, which are seen as small "drops", for both regimens. It should be noted the peculiar "elongated drop" shape of the bigger particles in the thin film grown with single-pulses, whose origins are still being determined. The thin films made by multi- pulse ablation have a higher density of particles that the single-pulse ablation regime, but these are smaller, the particles size are between 30 - 280 nm, with 50 nm of average size. While in single-pulse thin film the particle size is between 30 - 600 nm, with average size of 100 nm. A quantitative size distribution is unviable, due to irregular form of the particles in thin films made with single-pulse ablation.

The second stage of the comparison consisted in evaluating the optical transmittance of the films. These were done using the transmittance technique in the UV-VIS-NIR range. The measurements were performed with a lambda 40 Perkin Elmer spectrometer. Figure 7 shows the transmittance spectra of the thin films grown by both regimens. The films corresponding to 6000 and 3000 multi-pulses, with thicknesses of

65 nm and 34 nm respectively, have a significant absorption rate in the infrared region, especially between 800 - 1200 nm. This can be due to the vacancies of oxygen in the films as a result of the vacuum conditions used in the deposition [20] . These vacancies of oxygen can favor the inclusion of very small proportions of metallic titanium in the films, which would give rise to the absorption in the infrared-visible region. On the other hand, the films made with 6000 and 9000 multi-pulses, with 30 and 63 nm of thickness respectively, show an absorption close to zero in the infrared region of the spectrum. This could indicate that films grown with multi-pulses allow an improved conservation of the stoichiometry, as reported in [21] .

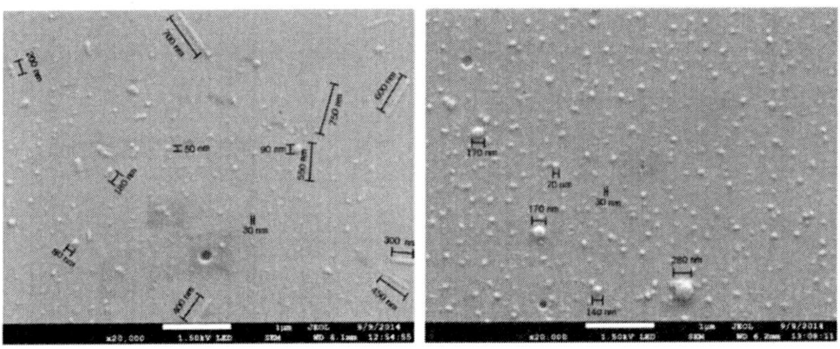

Figure 6. SEM micrography of TiO_2 thin film made with: (left) 3000 single-pulses and (right) 6000 multi-pulses.

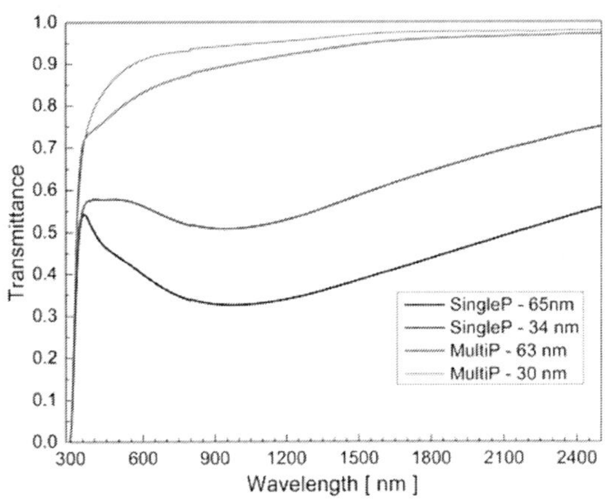

Figure 7. Transmitance spectra of TiO_2 thin films made with mono-pulse and multi-pulse laser ablation.

4. CONCLUSION

Single-pulsed ablation provided a 2:1 higher deposition rate than multi-pulsed ablation; consequently it can be stated that ablation in single-pulse regime is more efficient in terms of the extraction of material per pulse. However, for similar layer thicknesses, multi-pulsed ablation provided films that were not as rough and with particle size less. The films obtained with multi-pulses also have a transmittance greater level, which is desirable for most of the optical applications of TiO_2 thin films.

ACKNOWLEDGEMENTS

The authors thank CONACYT for financial support.

REFERENCES

1. Rout, S., Popovici, N., Dalui, S., Paramês, M.L., da Silva, R.C., Silvestre, A.J. and Conde, O. (2013) Phase Growth Control in Low Temperature PLD Co:TiO_2 Films by Pressure. Current Applied Physics, 13, 670-676.
2. Han, K. and Kim, J.H. (2012) Fabrication of TiO_2/SiO_2 Multilayer Film Structure by the Sol-Gel Process with Efficient Thermal Treatment Methods. Applied Surface Science, 263, 69-72.
3. Sivalingam, K., Shankar, P., Mani, G.K. and Rayappan, J.B.B. (2014) Solvent Volume Driven ZnO Nanopetals Thin Films: Spray Pyrolysis. Materials Letters, 134, 47-50.
4. He, C., Li, X.Z., Graham, N. and Wang, Y. (2006) Preparation of TiO_2/ITO and TiO_2/Ti Photoelectrodes by Magnetron Sputtering for Photocatalytic Application. Applied Catalysis A: General, 305, 54-63.
5. Kamata, K., Maruyama, K., Amano, S. and Fukazawa, H. (1990) Rapid Formation of TiO_2 Films by a Conventional CVD Method. Journal of Materials Science Letters, 9, 316-319.

6. Boyd, I.W. (1996) Thin Film Growth by Pulsed Laser Deposition. Ceramics International, 22, 429-434.
7. Dietsch, R., Holz, Th., Mai, H., Panzner, M. and Völlmar, S. (1995) Pulsed Laser Deposition (PLD)—An Advanced State for Technical Applications. Optical and Quantum Electronics, 27, 1385-1396.
8. Krebs, H.-U., Weisheit, M., Faupel, J., Süske, E., Scharf, T., Fuhse, C., Buback, M., et al. (2003) Pulsed Laser Deposition (PLD)—A Versatile Thin Film Technique. In: Kramer, B., Ed., Advances in Solid State Physics, Vol. 43, Springer, Berlin Heidelberg, 505-518.
9. García, T., de Posada, E., Ponce, L., Sánchez, J.L., Díaz, S., Pedrero, E. and Pereira, J.A.M. (2001) Textured Strontium Ferrite Thin Films Grown by PLD. Materials Letters, 49, 294-298.
10. Kitazawa, S.-I., Choi, Y. and Yamamoto, S. (2004). In Situ Optical Spectroscopy of PLD of Nano-Structured TiO_2. Vacuum, 74, 637-642.
11. He, X.L., Wu, J.H., Li, X.M., Gao, X.D., Zhao, L.L. and Wu, L.N. (2009) Synthesis and Properties of Silicon Dioxide Films Prepared by Pulsed Laser Deposition Using Ceramic SiO_2 Target. Applied Surface Science, 256, 231-234.
12. Zhao, J., Zhao, S.Z., Li, K., Kong, F. and Zhang, G. (2011) Optimization of Passively Q-Switched and Mode-Locked Laser with Cr4+:YAG Saturable Absorber. Optics Communications, 284, 1648-1651.
13. Poncea, L., Moreirab, L., Floresa, T., Arrontea, M., de Posadaa, E., Rodrígueza, E. and Hernandezb, L. (2009) Laser Induced Breakdown Spectroscopy; Advances in Resolution and Portability. Revista Cubana de Física, 27, 94-98,
14. Jedlinszki, N. and Galbács, G. (2011) An Evaluation of the Analytical Performance of Collinear Multi-Pulse Laser Induced Breakdown Spectroscopy. Microchemical Journal, 97, 255-263.
15. Koechner, W. (2006) Q-Switching. In: Koechner, W., Ed., Solid-State Laser Engineering, Vol. 1, Springer New, York, 488-533.

16. Radziemski, L. and Cremers, D. (2013) A Brief History of Laser-Induced Breakdown Spectroscopy: From the Concept of Atoms to LIBS 2012. Spectrochimica Acta Part B: Atomic Spectroscopy, 87, 3-10.
17. Martín, J.M.A. (2003) Láminas delgadas y recubrimientos: Preparación, propiedades y aplicaciones. CSIC, Madrid.
18. Ohring, M. (1992) The Materials Science of Thin Films. Academic Press, New Jersey.
19. Mitra, J., Paranjape, M., Raychaudhuri, A.K., Mathur, N.D. and Blamire, M.G. (2005) Temperature Dependence of Density of States Near the Fermi Level in a Strain-Free Epitaxial Film of the Hole-Doped Manganite La 0.7 Ca 0.3 MnO_3. Physical Review B: Condensed Matter and Materials Physics, 71, Article ID: 094426.
20. Chen, L.-C. (1994) Particulates Generated by Pulsed Laser Ablation. In: Hubler, G.K., Chrisey, D.B., et al., Eds., Pulsed Laser Deposition of Thin Films, John Wiley & Sons, Hoboken, 167-197.
21. Alvira, F.C., Ponce, L.V., Flores, T. and Peñaloza-Mendoza, Y. (2014) Pulsed Laser Deposition of PbTe in Monopulse and Multipulse Regime. Paper Presented at the Latin America Optics & Photonics Conference (LAOP), Cancun, Mexico. LM4A.29.

Chapter 11

ND:YAG Laser Pulses Ablation Threshold of Stainless Steel 304

Mitra Radmanesh, Amirkianoosh Kiani*

Department of Mechanical Engineering, University of New Brunswick, Fredericton, Canada

ABSTRACT

Laser processing and laser surface texturing in multiple fields have become a popular topic of study in recent decades. Understanding the principles behind the laser irradiation mechanism i an essential step in choosing the most effective process parameters. Through this study, the effects of power and pulse duration on the structure and surface pattern of stainless steel type 304 were examined, and optimized laser parameters were introduced for desired laser penetration and heat-affected areas on the surface. The analyzed sample was prepared by using variations of pulse durations and different pulsed energies. Looking at the trend of change of non-dimensional temperature along the surface, thickness, and center of the sample, the effects of pulse duration and intensity (corresponding to energy) were observed. Upon considering all the aspects of the irradiated spots, such as heat-affected area diameter, surface patterns, and penetration depth, the advantages and disadvantages of short and long pulse durations are mapped out clearly. Also, a new method to obtain the ablation threshold of stainless steel is introduced, and a thorough analytical solution is obtained.

Keywords Laser Material Processing, Laser Surface Treatment, Stainless Steel, Ablation Threshold

1. INTRODUCTION

In recent decades, lasers involving technologies and modeling of laser material interactions have gained notable attention in micro/nano manufacturing fields. Some of the most important applications of lasers are treatment and processing of variety of metals and alloys. Laser material processing involves high concentrated heating of solid targets using lasers to enable several types of structural changes, such as amorphization, hardening, and surface patterning, in a fast and precise manner, in micro, sub-micro, and nano scales [1] [2]. Lasers are able to deliver a range of very low to extremely high energy with extreme precision in dimension, spatial, and temporal distribution. In comparison to conventional methods, this laser approach is more flexible and reduces rocessing time, making it particularly suitable for both mass production and rapid prototyping and custom-scale manufacturing for a wide variety of pplications such as micro-welding, drilling, cutting, and heat treatment of etals and alloys [3] [4].

Laser processing can be applied in two categories based on the energy requirements: applications that require relatively low energy with limited structural and physical changes, such as annealing of semiconductors, and applications that call for high energy transforms for significant structural changes over a large volume, such as welding [5]. Energy transformation in any application involving lasers is fully dependent on the nature of the interacting material and its chemical bonding. In recent decades, there have been significant improvements, by development of ultra-short lasers, to increase the interaction time (pulse duration) between the laser and the material, hence reducing the effects on the bulk material [2] [6]. Despite these improvements, ultra-short laser systems are very expensive and cannot be utilized in industries. Therefore, depending on the nature of the application, less expensive laser systems with various pulse durations in the range of millisecond to microsecond that are widely available in manufacturing sectors are more popular among the manufacturers. A thorough understanding of the laser irradiation mechanism is required to utilize the more common lasers to their fullest efficiency.

In order to fully understand the mechanism behind laser irradiation as well as to increase the efficiency of the process, understanding the effects and controlling the process parameters is of utmost importance; therefore, choosing the most optimal pulse duration is essential to the final quality of a particular application [2] [7]. This study was conducted to identify the effects of key laser process parameters, such as power and pulse duration, in industrial laser systems on the structure and surface pattern of stainless steel 304, in order to develop an optimal process map for laser processing and surface texturing of materials. Also, a new method for calculation of ablation threshold has been introduced for stainless steel, which can be customized for other metals and alloys.

2. EXPERIMENTAL SETUP

Thin sheets of stainless steel type 304, with thickness of 0.5 mm were used for surface treatment in this study. The sample was irradiated with microsecond pulses to create the predetermined bullet-point patterns across its surface. Experiments were conducted using a pulsed Nd:YAG laser system with a peak power of 6 kW, and beam diameter of 20 um. For a wavelength of 1064 nm provided by this laser system, a coefficient of absorption of 0.3 was used for stainless steel type 304 throughout the calculations [8] [9].

The study was completed using variations of intensity of the laser system. The delivered intensity was controlled by reducing the voltage at 25 intervals starting at 400 V to 275 V. This trend was applied to three different pulse durations of 20 ms, 10 ms, 5 ms. **Table 1** visualizes the procedure of the experiment:

This experiment provided a very confined and controlled experimental environment for the required observations.

Upon completing the experiment, the effects of power and pulse duration on the surface structure, heat-affected zone on the surface, depth of penetration, and the created surface patterns were fully examined. After obtaining a thorough map of effects of process parameters, a series of optimal parameters are presented based on the observations

Table 1. Laser parameters of laser processing of stainless steel.

Laser Setup	Surface of the Sample (Stainless Steel 304)					
	Spot 1 400 V ($J/\mu m^2$)	Spot 2 375 V ($J/\mu m^2$)	Spot 3 350 V ($J/\mu m^2$)	Spot 4 325 V ($J/\mu m^2$)	Spot 5 300 V ($J/\mu m^2$)	Spot 6 275 V ($J/\mu m^2$)
Pulse Width 20 ms	0.244	-	0.186	0.159	0.132	0.107
Pulse Width 10 ms	0.158	0138	0.119	0.100	0.081	0.064
Pulse Width 5 ms	0.088	0.076	0.064	0.053	0.042	-

3. METHODOLOGY

Since laser processing mainly involves transformation hardening (which is dominantly a heat transfer process), examining the temperature distribution and modeling the thermal process from the initial stages of heating is an effective method to investigate the connection of process parameters with the desired outcome results [10] [11]. Ideally, the size and shape of the irradiated zone is required to analyze the obtained results. This information cannot be extracted directly by using the process parameters; thus, temperature history of the process must be developed to obtain the heat-affected profile of the work *piece* [12]. To develop the temperature history of the irradiated sample, the nondimensionalization method was introduced to track the changes due to each laser parameter, by calculating the non-dimensional temperature at the surface, depth, and center where the maximumnon-dimensionaltemperature occurs [13]-[15]. This method is a simple technique for tracking the trend ofchanges occurring in laser irradiation, and is an effective step in investigating the effects of laser parameters upon the heat-affected profile [4] [16].

To proceed with this method, a cylindrical symmetric flow without internal heat generation, and constant, isotropic material properties was assumed; from there, a transient heat conduction equation and a proper boundary condition, as shown in the following, were defined to develop the temperature history of the process to estimate the temperature at each point [17].

$$\frac{1}{r}\frac{\partial}{\partial r}\left(r\frac{\partial T}{\partial r}\right)+\frac{\partial^2 T}{\partial z^2}=\frac{1}{\alpha}\frac{\partial T}{\partial t} \tag{1}$$

where α is the thermal diffusivity, and is defined as $\alpha = k\ \rho Cp$, with k being the thermal conductivity, and ρ the density, and $p\ C$ the specific heat.

$$\left.\frac{\partial T}{\partial z}\right|_{z=0}=-\frac{1}{k}\frac{\eta P(t)}{\pi r_0^2}\exp\left(-r^2/r_0^2\right) \tag{2}$$

where P is defined as the peak power delivered, ⍰ is the fraction of energy absorbed by the surface, and r_0 is the beam radius (in this study the beam diameter is 20 um)

Taking into account the vast range of scale variation between the required parameters (for example, while temperature changing is in orders of thousands of degrees, pulse duration only changes in orders of millise- conds), it is more convenient to define a set of non-dimensional groups, and to simplify the obtained equations in terms of temperature, radius, location of the spot, and the required constant values [11] [17]. The customized non-dimensional variables are as follow:

$$R = r/r_0 \; ; \; Z = z/r_0 \; ; \; \theta = T/T_{ref} \; ; \; \tau = 4\alpha t/r_0^2$$

where, R is non-dimensional radius, Z is non-dimensional depth, ?□□□is non-dimensional temperature (It is assumed that reference temperature is the melting temperature of the material), and finally, ?□□□is non-dimen sional time for $t \le t_p$.

These non-dimensional variables are then used to simplify Equations (1) and (2). Equation (3) displays the simplified boundary equation [12]:

$$\left.\frac{\partial \theta}{\partial Z}\right|_{Z=0} = -Q\exp\left(-R^2\right)\left[u(\tau) - u\left(\tau - \tau_p\right)\right]$$

where Q is the non-dimensional power and is defined as: $Q = \eta P / k\pi r_0 T_{ref}$

Taking the Laplace transform of the redefined heat conduction equation, and solving the resultant differential equation, we can then calculate the integral form of the temperature. Hence, the maximum surface temperatures of various pulse durations can be obtained to show the non-dimensional temperature profiles under each condi- tion. As stated, the non-dimensional temperature in this study was tracked along the radius, thickness, and center of the work piece. Equations (4)-(6) define these temperature profiles [12] [17]:

$$\theta\left(R, Z = 0, \tau_p\right) \approx \frac{Q}{\sqrt{\pi}} \arctan\left(\sqrt{\tau_p}\right) \exp\left(-R^2\right) \tag{4}$$

$$\theta\left(R = 0, Z, \tau_p\right) \approx \frac{Q}{\sqrt{\pi}} \arctan\left(\sqrt{\tau_p}\right) - QZ \tag{5}$$

$$\theta(0, 0, \tau) = \frac{Q}{\sqrt{\pi}} \sqrt{\tau}, \quad \text{for } \tau < \tau_p \tag{6}$$

Comparing the temperature profiles of each case can consequently lead to estimating the most optimal process parameters for more efficient laser processing applications.

An important theory to be considered when working with non-dimensional temperatures is the behaviour of the temperature profile in cases of constant power and constant energy. When looking at a constant power case, a shorter pulse duration results in a lower non-dimensional temperature [16]. This is because when constant power is delivered by the laser system, a shorter irradiation ime results in less heat being accumulated, and a lower non-dimensional temperature is achieved. On the other hand, when considering a constant energy case, the non-dimensional temperature is higher with shorter pulse durations [16]. This can be explained by the fact that with a constant amount of energy, a shorter pulse duration allows the energy less time to spread along the work piece. In other words, energy is delivered in a more concentrated manner with shorter pulse durations; hence, higher non-dimensional temperatures resulted. With this behaviour in mind, experimental observations were examined to determine the most optimal process parameters [16].

4 RESULTS AND DISCUSSION

4.1 Heat Affected Zones and Keyhole Generation

Multiple bullet point patterns with stated properties are created upon the surface of the indicated stainless steel sample. **Figure 1** displays the optical images taken of the first four points of 20 ms, 10 ms, and 5 ms pulse dura- tions, respectively. As displayed, the size and surface texture of each spot changes based on the corresponding intensity and pulse duration.

Each irradiated spot displays different layers and patterns across its surface. This creation of different zones on the surface is due to the keyhole generation phenomenon that takes place during any laser surface texturing process [9] [18]. In **Figure 1**, each picture is almost consisted of three zones; **Figure 2** displays these zones and creation of keyhole in more details. **Figure 2(a)** presents the different zones created across the surface. The out- er area, which is indicated with the solid circle, marks the secondary heat-affected zone. The next layer shown with the dashed circle indicates the primary heat-affected zone (HAZ), and is the center of the penetration zone. Lastly the shaded area (molten area), which appears as waves on the surface of the irradiated spot, signifies the molten area caused by the accumulation of heat on the surface of the work piece [9] [19].

Figure 2(b) presents the location of keyhole generation with respect to the zones created across the irradiated spot. In this figure, the white triangle is the primary heat-affected zone, the shaded area is the molten area, and larger triangle signifies the secondary heat-affected zone. When processing materials using a laser, controlling the size of each of these layers is an essential factor in the precision and quality of the final results. Hence, it is essential to identify the effects of intensity and pulse duration on these layers [9] [19]. The keyhole is generated in the white triangle at the center of the irradiated spot, which corresponds to the primary heat-affected zone

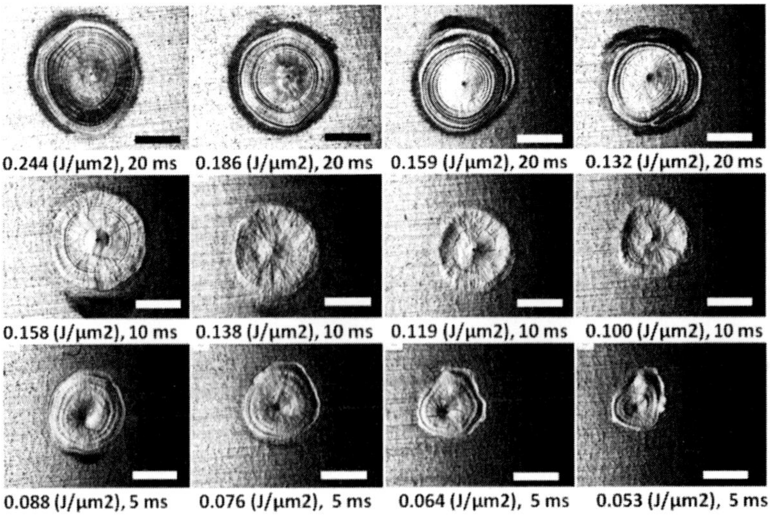

Figure 1. Optical images of irradiated spots (scale bar: 30 μm).

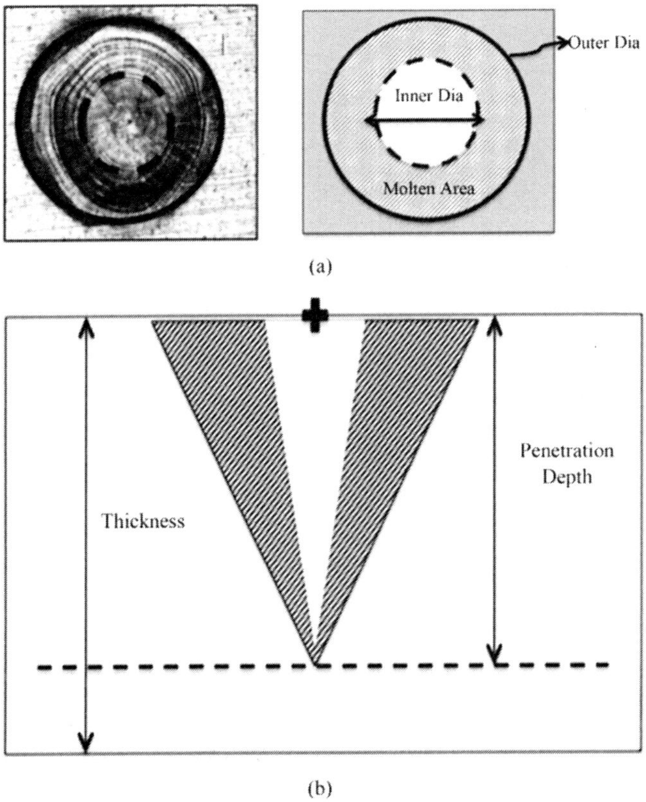

Figure 2. Surface zones and keyhole generation (scale 10 um).

4.2 Non-D Temperature Profile: Variation of Heat Affected Zones

The effects of power and pulse duration on the size of each indicated zone can be further examined by looking at the non-dimensional temperature profiles of various pulse durations across the corresponding outer diameters (primary heat-affected zone). **Figure 3** illustrates the variation of the outer diameter (as indicated in **Figure 2**) based on intensity values.

It is important to note that in this study intensity is obtained based on the energy delivered to the sample. As expected, the diameter of the heat-affected zone increases with an increase in intensity, hence an increase in the delivered energy. At a constant intensity (energy per

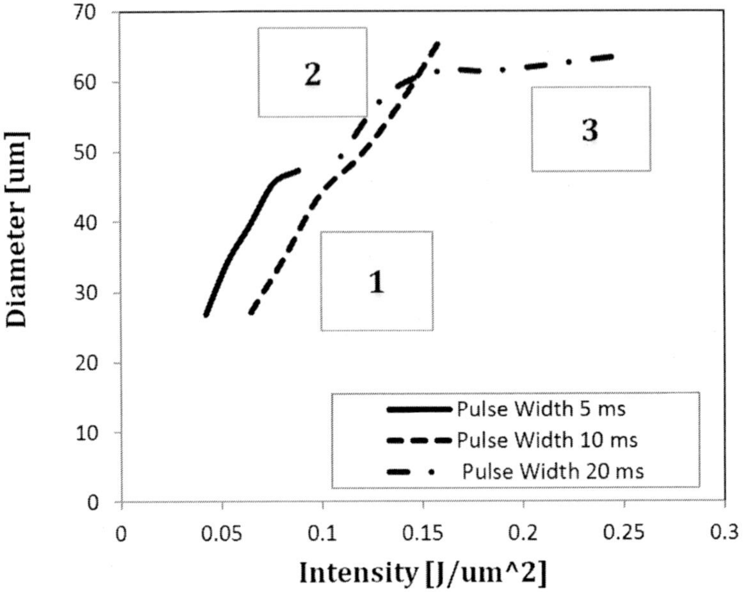

Figure 3. Outer diameter vs. intensity.

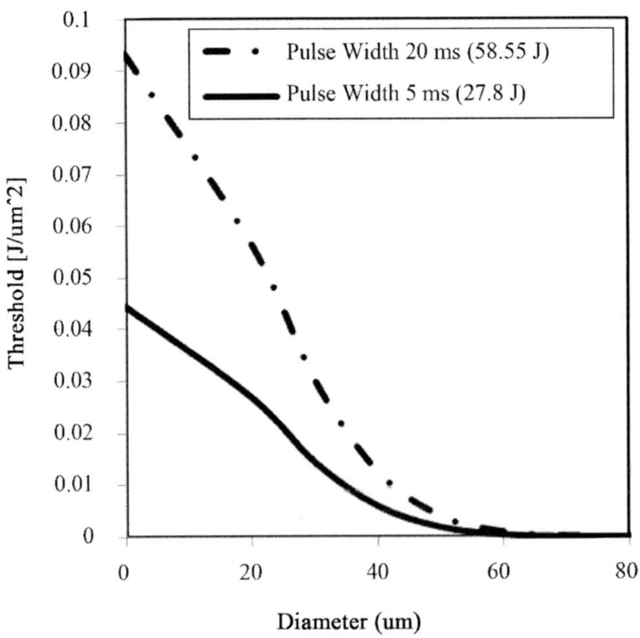

Figure 4. Computed results for ablation threshold (Equation (9)).

area), the shorter pulse duration results in a larger outer diameter. A case of a constant intensity limits the amount of available energy; therefore, when using shorter pulse duration, the constant energy will be delivered in a more concentrated manner, since the process is not long enough for the energy to spread across the material. This in turn results in a more heat accumulation, and therefore a larger heat-affected zone diameter is achieved using shorter pulse duration [16].

After examining the behaviour of the outer diameter, in accordance to intensity and pulse duration, exploring it at various key points is essential in understanding this behaviour. The important zones at which significant changes occur in the behaviour of outer diameter are indicated in **Figure 4**. Starting with Zone 1, the trend of changes is as expected, and as theory suggests, the outer diameter increases with an increase in intensity, and shorter pulse duration results in a larger diameter value at a constant intensity. However, when zone 2 is reached, it can be seen that the diameters start to have similar values, particularly for 10 ms and 20 ms pulse durations. This turn of events suggests that the effects of pulse duration on the heat-affected zone start to decrease as the intensity increases. This indicates that the temperature of the material is starting to reach the ablation temperature threshold, and that the material is reaching the saturation state. Laser ablation is generally defined as material removal in macroscopic scale from the surface of the work piece, due to a significant change of state, where the ablated amount transforms into the gas or plasma phase [20] [21]. In cases where laser radiation reaches a suffi- ciently high intensity, the surface of the material begins to ionize and transforms into a form of dense plasma. When the delivered energy by the laser is increased while the pulse duration is decreased, the accumulated heat will eventually exceed the breakdown threshold of the material. Thus, ablation is expected, and the size of heat-affected zone (HAZ) reaches its saturation point [21]. Looking at zone 3, it is clear the material has reached the saturation point, and the diameter remains constant despite the changes in intensity. Reaching the saturation point suggests the material has exceeded its ablation threshold at this point, and as the intensity increases the material starts evaporating;, consequently, the diameter of the heat-affected zone remains constant [22]. Deter- mining the threshold for ablation temperature is very essential in laser processing of materials.

4.3 Ablation Threshold of Stainless Steel

In this work, a method to develop the ablation threshold of stainless steel along with a thorough analytical pro- cedure is presented. This method can be further customized to accommodate the ablation thresholds of any other metals and alloys.

For lasers with Gaussian beam profile, if heat propagation effects are neglected, the feature size of the ablated area can be calculated by [23] [24]:

$$D^2 = 2r_O^2\left\{\ln \phi_O - \ln \phi_{th}\left(N_s\right)\right\} \tag{7}$$

where, rO is the radius of laser spot, $\varphi th(Ns)$ is the multi-pulse ablation threshold which depends both on the number of pulses and material and φO is the maximum laser flounce which is given by:

$$\phi_O = \frac{2E_{\text{pulse}}}{\pi r_O^2} \tag{8}$$

From Equations (7) and (8), the one pulse ablation threshold of the stainless steel can be calculated by:

$$\phi_{th}(1) = 2E_{\text{pulse}} \Big/ \pi \cdot r_O^2 \cdot \mathrm{e}^{0.5(D/r_O)^2}$$

Figure 4 shows the computed results for $\varphi th(1)$ at different pulse durations based on the analytical solution. As shown in the figure, the ablation threshold for one pulse decreases with decrease in the pulse duration. Also, the threshold nearly reaches zero at diameter of 60 um which is in a good agreement with our experimental re- sults shown in **Figure 3**. Depending on the application required, the intensity and pulse duration could be varied and controlled in or- der to reach or avoid the ablation threshold and achieve the desired results in any of the presented zones.

4.4 Relation of Maximum Non-D Temperature, Intensity, and Non-D Power

The trend of change in heat-affected zones can be further explained by looking at the relation between the maximum non-dimensional temperature, intensity, and non-dimensional power used with each indicated pulse durations. **Figure 5** presents this comparison.

Figure 5, the non-dimensional temperature, increases with an increase in both intensity and power. Thus, at higher intensity and power values, higher temperature is delivered to the materials, and a larger area is affected. This is in close agreement with the changes occurring in the outer diameter based on energy and intensity.

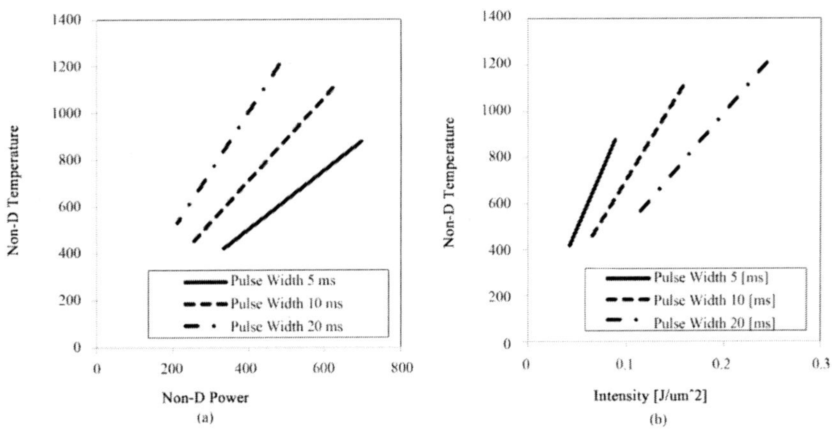

Figure 5. Max non-dtemperature vs. intensity & non-d power (Equations (4)-(6)).

Looking closely at graph 1), at a constant non-dimensional power, shorter pulse duration results in a lower non-dimensional temperature. However, in graph 2) the opposite is observed: at a constant intensity, the shorter pulse duration develops a higher non-dimensional temperature. This corresponds with the behaviour of the outer diameter at different pulse durations and intensity values presented in **Figure 3**. It is clear the higher non-di- mensional temperatures result in larger heat-affected areas, and at a constant intensity the shorter pulse duration results in both higher non-dimensional temperature and larger outer diameter. Shorter pulse duration causes the delivered energy to be more concentrated on the irradiated spot, and hence increases the accumulated heat [16].

4.5 Non-D Temperature Profile; Surface, and Depth

4.5.1 Effects of Laser Parameters on Surface Patterns

The next portion of the study mainly concentrated on the changes in non-dimensional temperature along the surface and depth of the work piece. According to **Figure 2**, both outer and inner diameters change relative to each other with intensity and pulse durations. To demonstrate this relativity, **Figure 6** displays the changes in the ratio of diameters with intensity. The outer to inner diameter ratio increases with an increase in intensity. This means that as more energy is delivered to the material, the heat-affected zone grows and inner diameter de- creases, resulting in a larger molten zone area. It should also be noted that looking at a constant intensity, the shorter pulse duration results in a larger diameter ratio, which is in a close agreement with the previous analyti- cal results.

In **Figure 2**, it was observed that the indicated molten area appears as a range of waves on the surface of the irradiated spot. When a material surface is irradiated using a pulsed laser system, several physical processes such as desorption, melting, vaporization, boiling, or even spontaneous evaporation (ablation) may take place that could result in formation of wave-like topography on the surface of the material [22]. Multiple theories present the reasons for these periodic structures appearing upon the surface of the irradiated material, but they have mainly been related to the capillary waves [22].

Surface pressure gradient could be an initiative for the formation of these structures. It is known that the pressure near the concave portion of a surface of a material is larger than on the convex parts. Upon melting of the surface material due to laser irradiation, this pressure difference could cause the molten liquid to move to- wards the convex regions of the surface, resulting in the creation of wave-like patterns [25]. Another possibility could be the non-homogenous deposition of energy that would initiate the motion of the molten material from hotter regions to colder areas. However, the key element that is common in both cases is the motion of the mol- ten material to increase the amplitude of the wave-like structures. This is contrary to the tendency of molten material for a wetting of its own solid, and is due to the existing surface forces acting on the liquid. This results in creation of the wave-like patterns on the surface of the material [26].

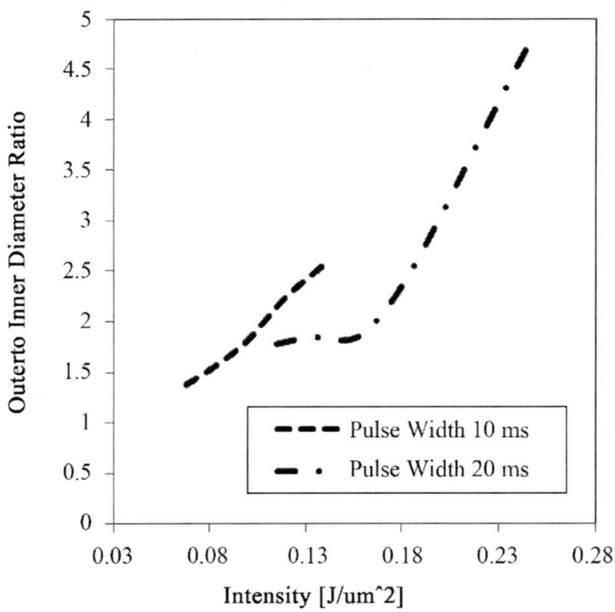

Figure 6. Outer to inner diameter ratio vs. intensity.

4.5.2 Non-D Temperature along the Surface

To further understand the relation among the outer and inner diameter due to the variation of laser parameters effects, the non-dimensional temperature along the surface is examined. **Figure 7** demonstrates this variation along the surface. The non-dimensional surface temperature is at its maximum at the center of the irradiated spot, gradually decreases along the radius of the spot, and eventually reaches a constant value at the room temperature. The three layers created on the surface of the irradiated spot can easily be shown along these curves. As indi- cated in the figure by the lines on the curves, the middle zone corresponds to the molten area shown in **Figure 2**.

As presented in the experimental results, at shorter pulse durations the secondary heat-affected zone (molten area) is significantly smaller than with the longer pulse durations; this can clearly be seen in **Figure 7**, as the solid line which is an indicator for pulse duration of 5 ms has the smallest part in the middle section of the curves compared to the dashed lines. A shorter pulse duration delivers the energy in a more concentrated manner; hence, shorter pulse durations involve rapid energy deposition and creation of vapour and plasma phases, and result in absence of molten area [25]. This reduces the effects of energy on the surrounding of the target area (heat accumulation). Therefore, the non-homogenous aspect of energy deposition is reduced, which causes the creation of wave-like patterns. It can be concluded that shorter pulse durations result in a better surface resolution, and smaller heat-affected zone [25].

4.5.3 Effects of Laser Parameters on Depth and Penetration & Non-D Temperature Profile at Depth

After examining the changes across the surface of the sample, it is essential to look at the effects of laser para- meters, particularly pulse duration, on depth and penetration of metals. **Figure 8** demonstrates the trend of change in the non-dimensional temperature along the thickness (depth) of the sample with regards to intensity. It also displays the variation of the heat-affected zone at the back of the sample to the outer diameter ratio with in- tensity.

In **Figure 8(a)**, the non-dimensional temperature increases as the intensity increases. This indicates that the penetration depth increases at higher intensities, which is in agreement with expectations: higher intensity deliv- ers higher energy to the material; hence the laser affects a larger amount of the material. As with previous cases, if we consider a case of constant intensity, it can be seen that the shorter pulse duration results in a higher non- dimensional temperature at the depth as well; hence a deeper penetration is achieved. This is further supported by looking at the heat-affected area on the back of the sample, which is displayed in **Figure 8(b)**; this graph presents the change in the heat-affected zone on the back of the sample as the intensity increases [12]. Back to outer diameter increases when increasing the intensity, which indicates that the heat-affected zone on the back of the samples increases with intensity as well. Therefore, in order to achieve a better surface resolution and a clearer and deeper penetration, shorter pulse durations should be used in the applications. However, with appli- cations that take place on the surface of the material, such as surface annealing of stainless steel, larger pulse durations should be used since they resulted in a larger heat-affected zone on the surface of the materials [12].

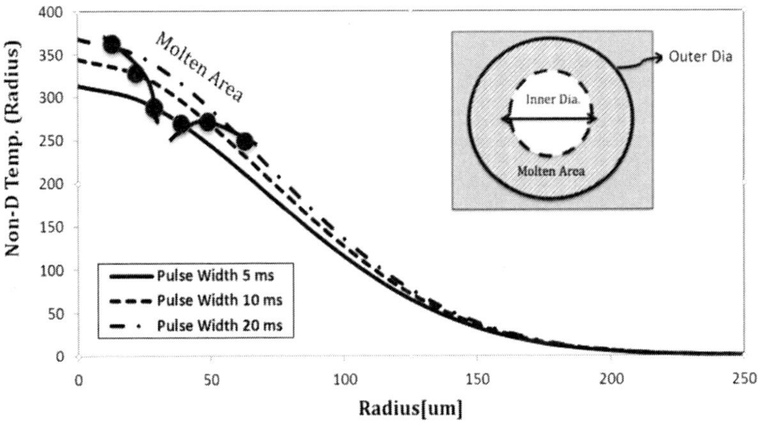

Figure 7. Non-dimensional temperature across the surface (Equation (4)).

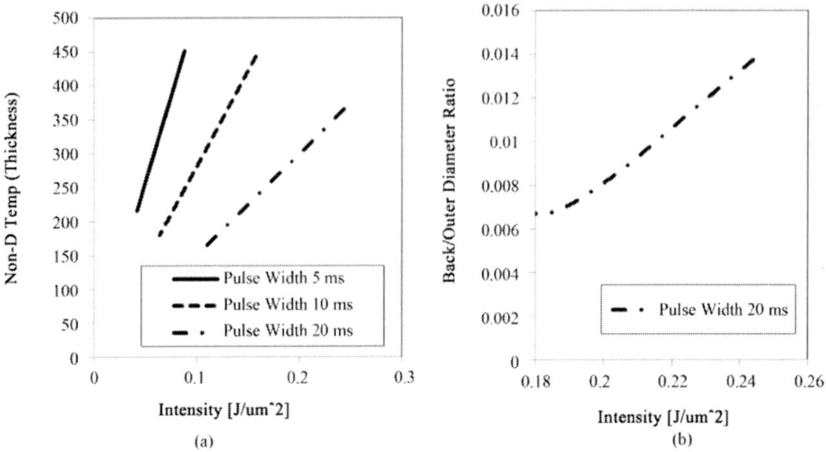

Figure 8. Non-D temperature at depth (Equation (5)) & back to outer diameter ratio vs. intensity.

5 CONCLUSION

A stainless steel type 304 sample was used to investigate the effects of pulse duration and power on the surface structure, patterns, and penetration, during laser irradiation. Also, a method for obtaining the ablation threshold of stainless steel for one pulse was introduced, which could be customized for any other metals and alloys. After a thorough analysis, it was observed that pulse duration was an essential factor in controlling the quality of the final product. In order to obtain a better surface resolution with minimum damage to the surrounding areas of the irradiated spot, shorter pulse durations should be used. For applications involving surfaces, larger pulse dura- tions are recommended due to their larger heat-affected zone diameters. Adjusting the intensity, that is, the energy delivered to the material, can control the depth of the penetration. Further, the introduced method for the ablation calculation showed that the ablation threshold for one pulse decreased with decrease in the pulse duration, and that the results were in very close agreement to the obtained experimental results.

REFERENCES

1. Majumdar, J. and Manna, I. (2011) Laser Material Processing. *International Materials Reviews*, **56**, 341-388.
2. Ahmmed, T., Yang, E.J., Ling, P., Servio B. and Kietzig, A. (2014) Introducing a New Optimization Tool for Femto- second Laser-Induced Surface Texturing on Titanium, Stainless Steel, Aluminum and Copper. *Optics and Lasers in Engineering*, **66**, 258-268.
3. Harooni, M., Carlson, B., Strohmeier, B.R. and Kovacevic, R. (2014) Pore Formation Mechanism and Its Mitigation in Laser Welding of AZ31B Magnesium Alloy in Lap Joint Configuration. *Materials & Design*, **58**, 265-276.
4. Brügmann, M.H. and Feurer, T. (2014) Comparative Theoretical Analysis of Continuous Wave Laser Cutting of Met- als at 1 and 10 μm Wavelength. *Applied Physics A*, **116**, 1353-1364.
5. Qiu, F., Uusitalo, J. and Kujanpaa, V. (2012) Laser Transformation Hardening of Carbon Steel: Microhardness Analy- sis on Microstructural Phases. *Surface Engineering*, **29**, 34-40.
6. Rizvi, N.H. (2003) Femtosecond Laser Micromachining: Current Status and Applications. *Riken Review*, 107-112.
7. Rulliere, C.E. (2005) Femtosecond Lasers; Principles and Experiments. Springer, Berlin.
8. Kuang, J.H., Hung, T.P., Lai, K., Hsu, C.M. and Lin, A.D. (2012) The Surface Absorption Coefficient of S304L Stain less Steel by Nd:YAG Micro-Pulse Laser. *Advanced Materials Research*, **472**, 2531-2534.
9. Kazemi, K. and Goldak, J.A. (2009) Numerical Simulation of Laser Full Penetration Welding. *Computational Mate- rials Science*, **44**, 841-849.
10. Tran, D.V., Lam, Y.C., Zheng, H.Y., Murukeshan, V.M., Chai, J.C. and Hardi, D.E. (2004) Femtosecond Laser. *Processing of Crystalline Silicon*, **19**, 11.

11. Yilbas, B.S. and Al-Aqeeli, N. (2009) Analytical Investigation into Laser Pulse Heating and Thermal Stresses. *Optics & Laser Technology*, **41**, 132-139.
12. Kiani, A., Venkatakrishnan, K. and Tan, B. (2013) Optimal Absorption Enhancement in 3D Silicon Oxide Nano-Sand- wich Type Solar Cell. *Optics Express*, **22**, A120-A131.
13. Ashby, M.F. and Easterling, K.E. (1984) The Transformation Hardening of Steel Surfaces by Laser Beams—I. Hypo- Eutectoid Steels. *Acta Metallurgica*, **32**, 1935-1937.
14. Li, W.B., Easterling, K.E. and Ashby, M.F. (1986) Laser Transformation Hardening of Steel—II. Hypereutectoid Steels. *Acta Metallurgica*, **34**, 1533-1543.
15. Ion, J.C., Scherclift, H.R. and Ashby, M.F. (1992) Diagrams for Laser Materials Processing. *Acta Metallurgica et Ma- terialia*, **40**, 1539-1551.
16. Kiani, A., Venkatakrishnan, K. and Tan, B. (2011) Enhancement of the Optical Absorption of Thin-Film of Amorpho- rized Silicon for Photovoltaic Energy Conversion. *Solar Energy*, **85**, 1817-1823.
17. Woodard, P. and Druden, J. (1998) Thermal Analysis of a Laser Pulse for Discrete Spot Surface Transformation Har- dening. *Journal of Applied Physics*, **85**, 2488-2496.
18. Tenner, F., Brock, C., Klampfi, F. and Schmidt, M. (2014) Analysis of the Correlation between Plasma Plume and Keyhole Behavior in Laser Metal Welding for the Modeling of the Keyhole Geometry. *Optics and Lasers in Engi- neering*, **64**, 32-41.
19. Hocheng, H., Tsai, H.Y., Jadhav, U.U., Wang, K.Y. and Lin, T.C. (2014) Laser Surface Patterning. *Comprehensive Materials Processing*, **9**, 75-113.
20. Linde, V.D. and Sokolowski-Tinten, K. (1999) The Physical Mechanisms of Short-Pulse Laser Ablation. *Applied Sur- face Science*, **154-155**, 1-10.

21. Nolte, S., Momma, C., Jacobs, H. and Tunnermann, A. (1997) Ablation of Metals by Ultrashort Laser Pulses. *Journal of the Optical Society of America B*, **14**, 2716-2722.
22. Dobrzanski, L.A. and Drygala, A. (2007) Laser Processing of Multicrystalline Silicon for Texturization of Solar Cells. *Material Processing Technology*, **191**, 228-231.
23. Venkatakrishnan, K., Stanley, P., Sivakumar, N.R., Tan, B. and Lim, L.E.N. (2003) Effect of Scanning Resolution and Fluence Fluctuation on Femtosecond Laser Ablation of Thin Films. *Applied Physics A*, **77**, 655-658.
24. Borowiec, A. and Haugen, H.K. (2004) Femtosecond Laser Micromachining of Grooves in Indium Phosphide. *Applied Physics A*, **79**, 521-529.
25. Erdogan, M., Oktem, B., Kalaycioglu, H. and Yavas, S. (2011) Texturing of Titanium (Ti6A14V) Medical Implant Surfaces with Mhz-Repetition-Rate Femtosecond and Picosecond Yb-Doped Fiber Lasers. *Optic Express*, **19**, 10986- 10996.
26. Aza, P.N., Fernandez-Pradas, J.M. and Serra, P. (2003) *In Vitro* Bioactivity of Laser Ablation Pseudowollastonite Coating. *Biomaterials*, **25**, 1983-1990.

Index